Oil and Gas Reserve Guidelines

Mike Zak　戴少武　郭齐军　著

中国石化出版社

图书在版编目（CIP）数据

石油天然气储量评估技术 = Oil and gas reserve guideline：英文 /（美）扎克（Zak, M.），戴少武，郭齐军著，— 北京：中国石化出版社，2014.8
ISBN 978-7-5114-2976-6

Ⅰ. ①石… Ⅱ. ①扎… ②戴… ③郭… Ⅲ. ①油气储量—油气资源评价—英文 Ⅵ. ①TE155

中国版本图书馆CIP数据核字（2014）第190101号

中国石化出版社出版发行
地址:北京市东城区安定门外大街58号
邮编:100011 电话:(010)84271850
读者服务部电话:(010)84289974
http://www.sinopec-press.com
E-mail:press@sinopec.com
北京柏力行彩印有限公司印刷
全国各地新华书店经销
*
700×1000毫米16开本14.75印张248千字
2014年8月第1版 2014年8月第1次印刷
定价：92.00元

《Oil and Gas Reserve Guidelines》
编写组

Mike Zak	戴少武	郭齐军	John Warner
胡允栋	王国鹏	苏永进	黄学斌
卢广钦	郭鸣黎	杜　霞	付　强
张社军	李　燕	李　冰	尚　峰

CONTENTS

Introduction

This is being written to compile what has been published and discussed concerning reserve reporting and to compare the various standards used to define and quantify oil and gas reserves. In light of recent developments and reserve write-downs, much of the emphasis of this paper will be on the United States Security and Exchange Commission (SEC) reserve definitions and how they apply to reserve reporting. It will also discuss the People's Republic of China (P.R.C.) reserve definitions, both the old and the recently published standards, the SEC definitions; again both the old and the December 31, 2008 revised SEC guidelines, and the SPE/WPC/AAPG/SPEE Petroleum Resources Management System (PRMS). This paper will not be all encompassing, nor will it necessarily give the same information as the SEC may give at some time in the future.

The Sarbanes-Oxley Act (SOX) is also applicable to the reserve evaluation process. Reserves are a major part of an oil and gas company's value, and so will be under more surveillance than prior to SOX. The added emphasis by SOX on transparency and the liabilities imposed on certain company officers will likely have oil and gas companies relying more on third party evaluations for their reserve estimates.

This paper is not written to provide legal advice or to provide reporting standards for reserves. Its intent is to provide the reader with a basis upon which to make decisions regarding reserves reporting and to organize the discussions available publicly. Each company should consult with their reserve estimator as to their unique reporting requirements.

Reserve estimates are prepared by oil and gas companies as a normal part of their business and may include not only estimates of reserve quantities, but also estimates of future producing rates, future net revenue, and the present value of the future net income.[1] The definitions used for the reserves reported should be identified and noted in the reserve report. The SEC reserve definitions are primarily concerned with reserve reporting in financial situations and as such focus on the proved category and are more stringent than those of the SPE/WPC. With the 2008 revision, the SEC allows, but does not require reporting probable or possible reserves as well as proved reserves. The purpose of the SEC definitions is to help mitigate the risk involved with oil and gas financial transactions. The Society of Petroleum Engineers/World Petroleum Congress (SPE/WPC) also allows the reporting of probable and possible reserves along with proved reserves. These categories of reserves involve more uncertainty and risk than proved reserves and are intended to be used by the company for planning and capital allocation.

The SPE/WPC revised their reporting document in March 2007. Four prior documents, the 1997 SPE/WPC Petroleum Reserves Guidelines, the 2000 SPE/WPC/AAPG Petroleum Resources Classification and Definitions, the 2001 SPE/WPC/AAPG Guidelines for the Evaluation of Petroleum Reserves & Resources and the 2005 APE/WPC/AAPG Glossary of Terms were combined into one document, the SPE/WPC/AAPG/SPEE Petroleum Resources Management System (PRMS). The new system has classification modifiers. The terms developed and undeveloped can be applied to proved, probable and possible reserves. Reserves, Contingent Resources and Resources can be modified by project maturity. Contingent Resources can be divided into marginal economic and sub-marginal economic. The PRMS applies to conventional as well as unconventional resources. The PRMS uses forecast conditions as the base case, but does allow the use of constant conditions.

The SEC revised its reserve reporting guidelines with new rules December 31,

1 Auditing Standards for Reserves, SPE, June 2001. "Estimates of Reserve Information are made by or for Entities as a part of their normal business practices. Such Reserve Information typically may include, among other things, estimates of (i) the reserve quantities, (ii) the future producing rates from such reserves, (iii) the future net revenue from such reserves and (iv) the present value of such future net revenue. The exact type and extent of Reserve Information must necessarily take into account the purpose for which such Reserve Information is being prepared and, correspondingly, statutory and regulatory provisions, if any, that are applicable to such intended use of the Reserve Information."

2008, which are effective January 1, 2010. This is the first major revision of the SEC reporting guidelines since they were initially adopted in 1978 and 1982. The SEC is attempting to align their rules with the PRMS guidelines and make them more compatible with current industry practices. The final rule is in 74 Fed. Reg. 2157 published on January 14, 2009.[2] The new rules should help the market to better evaluate the various oil and gas companies.

The December 31, 2008 SEC guidelines change the way prices are calculated to a 12 month average price, thus ending the prior use of the single day price at the end of the year. The SEC definition of proved oil and gas reserves has changed and now allows for the inclusion of various non-traditional resources, such as bitumen and oil and gas shale, as reserves. One difference between the PRMS and the SEC definitions is the use of the term "economic producibility" instead of "commerciality" in the definition of reserves. The new guidelines change the reporting of proved undeveloped reserves and require disclosure of progress toward development of those reserves. They allow, but do not require, the reporting of probable and possible reserves. The SEC has broadened the types of technology which can be used to establish reserves and reserve categories. The terms "reasonable certainty", "analogous reservoir" and "bitumen" have been defined by the SEC. The SEC guidelines will require disclosure of the qualifications of the person responsible for the final reserve estimate. The SEC will also require disclosure of the reserves in each foreign country having 15% of more of the company's reserves.

Reserve estimates are prepared for a variety of reasons and thus are based on a variety of definitions and guidelines. Primarily, reserve definitions provide a basis to quantify risk and uncertainty. Reserve estimates are required for oil and gas companies that filing with the SEC. Reserve estimates are used by oil and gas companies to plan their spending and evaluate which projects receive how much of the budget. Reserve reports are required by banks and other lending

---

2 "SEC Issues Final Rule on Modernization of Oil and Gas Reporting"
Fulbright Briefing, Robert S. Ballentine, Daniel M. McClure, Laura Ann Smith and Harva R. Dockery, January 2009.

institutions to provide security for loans meant for drilling or development of properties. Reserve estimates are used to determine the value of acquisitions and divestitures. Investors use reserve reports to determine the whether or not to participate in a project or invest in a company. Reserve reports are also used in litigation involving mineral interests. The user of the report should identify the guidelines used to prepare the report before relying on its conclusions.

China began allowing outside companies to explore for oil and gas in 1979 in the South China Sea, but the results were disappointing. In 1994, China opened the East China Sea and Bohia Bay. The discoveries in these two new areas are larger, but the geology is such that major discoveries and are not likely to be made. The sediments in China's basins are primarily fluvial rather than marine. The sands are discontinuous and the reservoirs are not as prolific as in some other basins. In 1985, China allowed the first outside onshore exploration in rather poor areas in the south and eastern parts of China. In 1994, exploration was finally allowed in a part of the Tariam Basin, but again the results were disappointing.

The rules affecting minerals in China were codified under the Mineral Resources Law of the PRC, March 19, 1986 and amended January 1, 1997.[3] The PRC reserve definitions have recently been rewritten to be more in line with those of the SPE/WPC. Under the new law foreign enterprises have the same exploration rights as domestic companies. Exploration permits are granted for seven years for oil and gas blocks. The basic exploration block is about 848 acres and is based on one minute of longitude and one minute of latitude. The maximum number of blocks is forty for an oil and gas exploration permit. The concessions run for three years, but can be renewed for two year periods if a commercial discovery is made.[4]

There are three state owned oil companies in China – CNPC, CNOOC, and SINOPEC. CNPC is the largest and operates mainly onshore along with SINOPEC. CNOOC operates primarily offshore, and is a partner with SINOPEC

3 Chinese Mining Law Overview, W. L. MacBride, Jr, Wang Bei, 2001.

4 Id.

in some offshore properties. All three of the Chinese oil companies are registered on the New York Stock Exchange and as such report reserves using SEC reserve definitions. Although the PRC reserve definitions are now similar to the Society of Petroleum Engineers/World Petroleum Congress (SPE/WPC) definitions and are moving more in that direction, there are still differences between them and the SEC reserve definitions. One of the purposes of this article is to help understand and explain those differences.

Who Defines Reserves?

The two primary sets of reserve definitions have been promulgated by the SEC and the SPE/WPC. Although there are other reserve definitions in use, they are not as widely used. There are efforts underway, however, to establish a set of definitions modeled around the SPE/WPC definitions that will be universally accepted. The SEC reserve definitions are required for the standardization of financial reporting by oil and gas companies, both domestic and foreign, registered on the stock exchanges of the United States. The purpose of the SPE/WPC reserve definitions, on the other hand, is to standardize reserves reporting by individual companies and countries.

Security and Exchange Commission (SEC)

The SEC's role is to protect investors and insure consistency in the financial markets. With the advent of Sarbanes-Oxley, the SEC is more motivated to insure full and complete disclosure to the investing public of a company's finances as well as transparency of the company's operations. The SEC definition of proved reserves is intended to comply with the SEC's stated mission to provide a consistent basis for reporting reserves among public oil and gas companies.

The SEC definitions were written at a time of long term gas contracts, more federal regulations of gas was sold on the interstate markets, and more stable prices. During that period of time, the petroleum industry was also more focused on the domestic arena rather than the international.

Financial Accounting and Standards Board (FASB)

The Financial Accounting Standards Board establishes standards for financial accounting and reporting. FASB69 establishes procedures for reporting reserves and costs and the year end pricing standard used in oil and gas reports used for documents submitted to the SEC.

Society of Petroleum Engineers (SPE)

The Society of Petroleum Engineers is an organization of professionals who work in the petroleum industry. Their stated mission is to collect and disseminate technical information about the exploration, development and production of petroleum resources for the benefit of the public.

The SPE originally promulgated reserve definitions in 1964, with revisions in 1981 and 1987. The WPC wrote its reserve definitions independently in 1983.

The SPE and WPC definitions were combined in 1997 for definitions of proved, probable and possible reserves (See Appendix D). They were written to provide a standardized set of definitions for the petroleum industry.

The SPE is still working to standardize reserve definitions worldwide. They monitor the activities of other groups and recommend revisions to the definitions as necessary. They have a committee to continually monitor revisions others are making to their reserve definitions.

The SPE has addressed standards for reserve audits and they are attached as Appendix E. They discuss compliance with reserve definitions, the qualifications of the reserve auditors, the standards of objectivity and independence, the standards to use when estimating or auditing reserves.

World Petroleum Congress (WPC)

The World Petroleum Congress is an organization whose stated purpose is to promote the management of petroleum resources worldwide for the benefit of mankind.

The World Petroleum Congress consists of 61 countries which represent over 90% of the major oil and gas producing and consuming countries in the world. It is involved in trying to establish a consistent worldwide set of reserve definitions and reporting standards. In 1987 they combined their definitions with those of the SPE.

American Association of Petroleum Geologists (AAPG)

The AAPG is an organization of geologists which is concerned with the science of petroleum geology.

In 2000, the SPE and WPC together with the AAPG published resource definitions, further expanding the definitions into Contingent and Prospective Resources (See Appendix G). These definitions address reserves that are not currently economic or technically feasible to develop and reserves that have not been discovered.

Society of Petroleum Evaluation Engineers (SPEE)

The SPEE is an organization of professionals whose primary focus is reserve estimation.

The SPEE sponsors annual forums designed to help the industry develop a better understanding of the SEC reserve definitions.

International Energy Agency (IEA)

The International Energy Agency was created in 1974 and consists of 26 countries in Europe including the United States. They are committed to joint efforts to meet supply emergencies, and to assist in the integration of energy policies. It is also trying to establish a consistent worldwide set of reserve reporting definitions and standards.

International Accounting and Standards Board (IASB)

The International Accounting Standards Board is equivalent to the FASB in the United States. They are concerned with a standardized set of reserve definitions.

United Nations (UN)

The UN is currently working to merge the UNFC classification and the SPE/WPC/AAPG classifications and establish a standardize system for coal, uranium and petroleum.

Energy Information Agency (EIA)

The Energy Information Agency of the U.S. Department of Energy uses the SPE/WPC reserve definitions. They collect data and make forecasts and analysis of petroleum information.

Alberta Securities Exchange

The Alberta Securities Exchange updated its reserve definitions in 2002 with NI 51-101. The Canadian Institute of Mining and Petroleum has provided reserve definitions for proved, probable and possible reserves. The Alberta Securities Exchange is equivalent to the SEC in the United States. Alberta Securities Exchange requires the use of outside auditors for companies under a certain size.

China

The Ministry of Geology and Mineral Resources was established by China in 1952. China's oil in place estimate was first made in 1953 in the Yumen Area of the Gansu Province. In 1998, the Ministry of Land and Resources was created with one of its major functions being reserves management.

In 1977, China through the Ministry of Petroleum issued guidelines for geologic reserve calculations. A criterion for oil and gas estimates was set up in 1982 by the Research Institute of Petroleum E&D. In 1984, the China National Reserves Committee began studying the criteria for oil and gas estimates.

China in 1988 published standards for reserve reporting through the National Standard Bureau which are included in the P.R.C. – Petroleum Reserve Standard.

Oil and Gas Reserve Guidelines

The different classifications of petroleum reserves are defined. The Petroleum Reserve Standard also explains the methods for estimating reserves. Since 2000, the Ministry of Land and Resource has been revising the reserve standards to make them more in line with the SPE/UNFC definitions. The new definitions are expected to be out shortly. [5]

In China, reserves are estimated by local experts and audited by government appointed evaluators. Reserves are estimated by the Local Project Teams and approved by the Local Reserves Management of the Regional E&P Company and Senior Management. The Ministry of Land and Resources, in the Petroleum Reserves Office, audits the proved (P-1) reserves after approval by senior management of the regional exploration and production company.

---

5 United Nations Economic and Social Council 21 Feb., 2001.
Economic Commission For Europe, Committee On Sustainable Energy; Practical Application Of The United Nations Framework, Classification For Reserves/Resources In China "5. Revision of the classification standards for oil and gas.
The current code for oil and gas classification was adopted in 1988. Then, there were a few proposals, since its promulgation, to revise it. Fortunately, the newly-developed *Classification for Resources/Reserves of Solid Fuels and Mineral Commodities* came out in a timely way and it acted as a catalytic promoter to revise the code for oil and gas classification. The revision work, using the definitions and classification criteria adopted by WPC and SPE for reference, has been almost completed up to date. The code could be summarized by the following features:
· The Stages of exploration are sub-divided into regional exploration, pre-exploration and evaluation exploration parts;
· The Degrees of economic viability are sub-divided into economic, sub-economic and intrinsic economic categories;
· Classify respectively at levels of geological resources, geological reserves and extractable reserves.
Geological resources are classified according to varying stages of exploration. Geological reserves are classified depending on both stages of exploration that the whole deposit is at and levels of geological confidence. Extractable reserves are classified depending on both levels of confidence with regard to the each part of the oil and gas deposits and application degrees of the technology by which the recovery is raised.

Major SEC Rule Changes of 2009

The new SEC rules entitled "Modernization of Oil and Gas Reporting" were published in the Federal register on January 14, 2009. The rules include revised definitions more in line with the SPE PRMS definitions of September 2007. [6]

The major changes from the new rules include the use of average an annual price based on first day of month prices instead of the prior required use of the one day year-end price. The new rules allow the use of technologies demonstrated in practice to be reliable, instead of listing certain specific technologies. The new rules recognize nontraditional resources such bitumen from oil sands and synthetic oil from oil shale and coal. The new rules allow the optional disclosure of probable and possible reserves. They've changed the criteria for proved undeveloped reserves beyond a one location offsets by replacing the word "certainty" in the definition by the words "reasonable certainty". [7]

The new rules consist of two major parts one the revised definitions and which amend Regulation S-X, Part 210 Rule 4-10 and the new disclosure records. If the new definitions are listed on pages 2190 to 2192 of the Federal register. The disclosure requirements were noted on pages 2193 to 2196 of the Federal register. [8]

In the past many companies indicated to the SEC that using one day year-end prices was not realistic. Under the new rules, the SEC replaced using one day year-end prices with prices determined from the arithmatic average of the price on the first day of the month for each month within the filer's fiscal year. [9]

The new rules also allow voluntary disclosure of probable possible reserves.

---

6 Modernization of the SEC Oil and Gas Reporting Requirements, Lee, W. J., SPE 123793, p.1.

7 Id.

8 Id.

9 Modernization of the SEC Oil and Gas Reporting Requirements, Lee, W. J., SPE 123793, p.4.

Oil and Gas Reserve Guidelines

Disclosure of proved reserves is still required. The term probable reserves as used by the SEC include reserves which together with proved reserves are more likely than not to be produced. In probabilistic terms the probability must be at least 50% the volumes recovered will equal or exceed the estimated of proved plus probable reserves.[10] The SEC also included a definition of resources in the new rules; however it does not allow disclosure of non reserve volumes. Resources are considered non reserves.

The SEC has relaxed the rules for proved undeveloped reserves. In the past, wells beyond one location offsets needed to satisfy the criteria of "certainty" of reservoir continuity. Under the new definitions, the SEC replaces the term "certainty" by the term "reasonable certainty".

The SEC will now allow the disclosure of certain run nontraditional resources as reserves. Examples of these include bitumen from oil shale, synthetic oil and gas from mined oil shale and coal, and gas hydrates.[11] Under prior SEC regulations, oil and gas produced from mined resources had to be disclosed as mineral resources not oil and gas reserves.[12]

The SEC now allows the use of reliable technology to establish gas reserves. In the past the SEC allowed only certain specified technologies such as the flow test or production tests. The new rules do not specify the technologies that can be used. The filer however has the burden to show that the technology is reasonable and reliable and has been demonstrated in practice to provide repeatable inconsistent results. Reliable technology can include proprietary technology the details of which are not required to be disclosed except to the SEC staff.

---

10 Modernization of the SEC Oil and Gas Reporting Requirements, Lee, W. J., SPE 123793, p.5.

11 Id.

12 Id.

SPE Reserve Classification

The SPE Petroleum Resources Management System classifies reserves and resources based on risk and uncertainty.

TOTAL PETROLEUM INITALLY-IN-PLACE (PIIP)
DISCOVERED PIIP
COMMERCIAL
SUB-COMMERCIAL
UNDISCOVERED PIIP
PRODUCTION
RESERVES
1P
2P
3P
Proved
Probable
Possible
CONTINGENT RESOURCES
1C
2C
3C
UNRECOVERABLE
PROSPECTIVE RESOURCES
Low Estimate
Best Estimate
High Estimate
UNRECOVERABLE
Increasing Chance of Commerciality
Range of Uncertainty
Not to scale

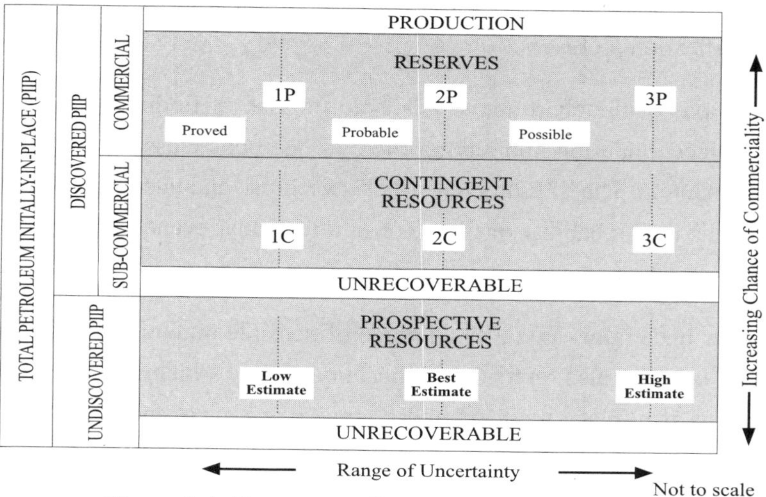

Figure 1-1: Resources Classification Framework.

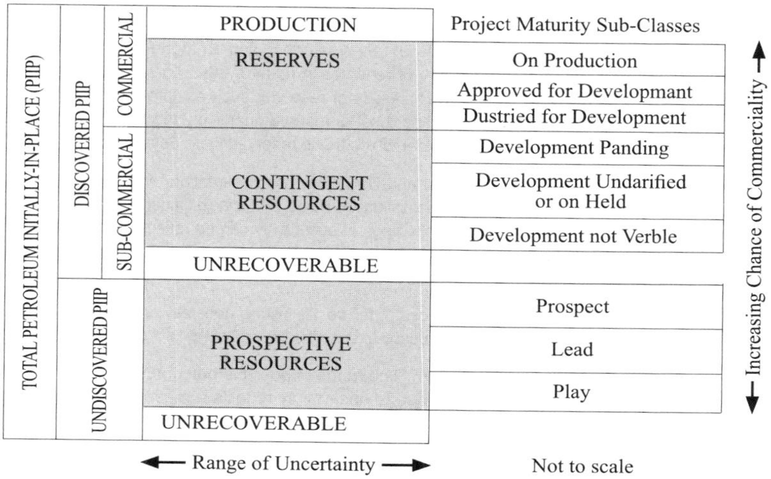

TOTAL PETROLEUM INITALLY-IN-PLACE (PIIP)
DISCOVERED PIIP
COMMERCIAL
SUB-COMMERCIAL
UNDISCOVERED PIIP
PRODUCTION
RESERVES
CONTINGENT RESOURCES
UNRECOVERABLE
PROSPECTIVE RESOURCES
UNRECOVERABLE
Project Maturity Sub-Classes
On Production
Approved for Developmant
Dustried for Development
Development Panding
Development Undarified or on Held
Development not Verble
Prospect
Lead
Play
Increasing Chance of Commerciality
Range of Uncertainty
Not to scale

Figure 2-1: Sub-Classes based on project Maturity.

SPE Petroleum Resources Management System

The SPE guideline is that reserves must be categorized according to the level of certainty that they will be recovered.[13] The range of uncertainty is determined by how much data we have to make the estimate and how certain we are of our assumptions.[14] As more data becomes available, the degree of uncertainty becomes less. Under the SPE guidelines, as the uncertainty decreases, the reserve classification increases.

The vertical axis of the chart above relates to the risk of finding a commercial hydrocarbon accumulation and having the hydrocarbons classified as reserves rather than resources. The SPE differentiates between risk and uncertainty. The SPE says that risk is the probability of failure or of a particular event happening.[15] The horizontal axis relates to the uncertainty of the reserve estimate.

The SPE says uncertainty involves a "range of possible outcomes in a series of estimates".[16] The SPE also refers to technical uncertainty which is the uncertainty due to a range of in-place numbers and a range of recovery factors.[17] When an estimate of the hydrocarbon in-place is made, there can be a series of values.

13 SPE Estimating and Auditing Standards for Reserves Sec 5.8 "Reserves must be classified according to the level of certainty that they will be recovered. To guide the classification of reserves, Reserves Definitions have been promulgated by various regulatory bodies and professional organizations throughout the world. Most sets of Reserves Definitions allow for different classes of reserves depending on the level of certainty associated with the reserves estimate. The highest category of reserves in many systems are "proved reserves," which require the highest degree of confidence. Lower classes of reserves such as "probable" or "possible" imply decreasing standards of certainty. When presenting a set of reserve quantities, the Reserve Estimator should always identify the set of Reserves Definitions under which those reserves were determined."

14 SPE Petroleum Resources Classification System and Definitions "Any estimation of resource quantities for an accumulation is subject to both technical and commercial uncertainties, and should, in general, be quoted as a range. In the case of reserves, and where appropriate, this range of uncertainty can be reflected in estimates for Proved Reserves (1P), Proved plus Probable Reserves (2P) and Proved plus Probable plus Possible Reserves (3P) scenarios. For other resource categories, the terms Low Estimate, Best Estimate and High Estimate are recommended."

15 SPE Glossary of Terms Used "The probability of loss or failure. As 'risk' is generally associated with the negative outcome, the term 'chance' is preferred for general usage to describe the probability of a discrete event occurring."

16 SPE Petroleum Resources Management System "The range of possible outcomes in a series of estimates. For recoverable resource assessments, the range of uncertainty reflects a reasonable range of estimated potentially recoverable quantities for an individual accumulation or a project. (See also Probability.)"

17 SPE Petroleum Resources Management System "Indication of the varying degrees of uncertainty in estimates of recoverable quantities influenced by range of potential in-place hydrocarbon resources within the reservoir and the range of the recovery efficiency of the recovery project being applied."

Also, within the range of in-place values, there is a range of recovery factors which can be assigned. The SPE suggests it useful to consider the recoverable quantities independently of the risk of commercial production. [18]

The SPE notes that reserve estimates will be revised as additional data becomes available.[19] In general, the more data available, the less uncertainty in the estimate. The SEC suggests the revisions for proved reserves should be in the upward direction.

Different categories of reserves are used for different purposes.[20] Most financial reports contain proved reserves. Some regulatory bodies and financial institutions also use probable reserves along with the proved. With the December 31, 2008 revision, the SEC allows companies to report probable and possible reserves, but does not require it.

The SPE definitions require proved reserves have a reasonable certainty of being recovered. Probable reserves are, in general, more likely than not to be recovered. And, possible reserves are less likely to be recovered than probable reserves.[21]

---

18 SPE Petroleum Resources Management System "While there may be significant risk that sub-commercial and undiscovered accumulations will not achieve commercial production, it useful to consider the range of potentially recoverable quantities independently of such a risk or consideration of the resource class to which the quantities will be assigned."

19 SPE Petroleum Resources Management System "Reserves estimates will generally be revised as additional geologic or engineering data becomes available or as economic conditions change."

20 SPE Estimating and Auditing Standards for Reserves Sec 5.8 "Different grades of reserves are used for different purposes. In general, proved reserves are used for financial reporting and lending, where the need for certainty is the greatest. However in order to make intelligent business decisions in activities such as prioritization of capital spending and property acquisitions, it is also important to recognize and quantify the amount of probable and possible reserves."

21 SPE Estimating and Auditing Standards for Reserves Sec 5.8 "The SPE/WPC (1997) definitions contains a general requirement that proved reserves have a "reasonable certainty" of being recovered. Other, more specific, criteria must also be met for reserves to be classified as proved. The definition for probable reserves is less stringent, requiring that a general test of "more likely than not" be satisfied. Possible reserves are those unproved reserves which analysis of geological and engineering data suggests are less likely to be recoverable than probable reserves."

Resources

The concept of resources and reserves was identified by McKelvy in 1972 and discussed by the AAPG together with the SPE/WPC in 2000. This model should be understood to fully understand the idea of reserves, as reserves are part of the overall classification of petroleum resources.

Resources refer to the total estimated volume of recoverable hydrocarbon, both discovered and undiscovered and are part of the Total Petroleum-Initially-In-place. Reserves, on the other hand, represent volumes economically recoverable from known accumulations as of a certain point in time. The main factors for classifying a hydrocarbon accumulation as a resource or as reserves are the maturity of the data acquisition and the commerciality of the project. The more mature the data acquisition and the more information available about the accumulation and commerciality, the farther it progresses along the line from resource to proven reserves.

The term resource is generally used to include all quantities of petroleum, including those not yet discovered and those already produced. The term "reserves" describes the part of the resource package from a certain date forward, meeting certain conditions and which is discovered.

SPE

The SPE defines resources as follows:

The term "resources" as used herein is intended to encompass all quantities of petroleum (recoverable and unrecoverable) naturally occurring on or within the Earth's crust, discovered and undiscovered, plus those quantities already produced. Further, it includes all types of petroleum whether currently considered "conventional" or "unconventional" (see Total Petroleum Initially-in-Place). (In basin potential studies, it may be referred to as Total Resource Base or Hydrocarbon Endowment.)[22]

22 SPE/WPC Petroleum Resources Management System; March 2007, Sec. 1.1.

The SPE subdivides Resources into categories to indicate uncertainty differences: "Subdivisions of estimates of resources to be recovered by a project(s) to indicate the associated degrees of uncertainty. Categories reflect uncertainties in the total petroleum remaining within the accumulation (in-place resources), that portion of the in-place petroleum that can be recovered by applying a defined development project or projects, and variations in the conditions that may impact commercial development (e.g., market availability, contractual changes)" [23]

The 2007 SPE-PRMS divides resources into two categories; either Contingent or Prospective. Prospective Resources have not yet been discovered. Contingent Resources have been discovered, but are not commercial. They may lack a market or be uneconomic using current technology or with current costs and prices. If a project has not been approved, the volumes are considered resources and not reserves.

To show commerciality under the PRMS, the project must have a reasonable timetable for development and that facilities are or will be available to produce the resources. There has to be a market or a reasonable expectation one will be available when the project is ready to produce. Governmental approvals need to be in place, or there must be a reasonable expectation they will be granted. [24]

The PRMS defines contingent resources as: "Those quantities of petroleum estimated, as of a given date, to be potentially recoverable from known accumulations by application of development projects, but which are not currently considered to be commercially recoverable due to one or more contingencies." [25] Contingent Resources may include projects which have no funding and volumes beyond the life of the concession or the sales contract. The technology needed to produce the volumes may still be under development. The project may be in the early stages of development and the commerciality not yet established.

---

23 SPE/WPC Petroleum Resources Management System; March 2007, Sec. 2.2.

24 SPE/WPC Petroleum Resources Management System; March 2007, p.6.

25 SPE/WPC Petroleum Resources Management System; March 2007, p.25.

Contingent Resources are generally denoted as C1, C2 and C3, with C1 having the lowest risk and C3 has the highest risk. 1C relates to the Probabilistic P90, 2C relates to the Probabilistic P50 and denotes C1 plus C2 volumes, and 3C relates to the Probabilistic P10 and denotes C1 + C2 + C3 volumes.

Since the Contingent classification is based on the lack of commerciality, one would expect the volumes to move to the corresponding reserve category when conditions are such they can be considered reserves. For example, one would expect C1 resources to become P1 reserves after the commerciality constraint is satisfied. Similar transfers would be expected of C2 to P2 and C3 to P3.

If part of a project is not commercial or not funded, the volumes which are commercial may be considered reserves, while the volumes which are not, would be resources. For example, if X locations are required to fully develop a discovery, but funding is expected for only 1/3X, then 1/3X locations and the associated volumes can be classified as reserves and 2/3X locations and associated volumes are classified as resources. The locations for which there is no funding should not be considered as probable or possible reserves, but rather as resources.

The PRMS defines Prospective Resources as: "Those quantities of petroleum which are estimated, as of a given date, to be potentially recoverable from undiscovered accumulations."[26] This can include undrilled prospects or even leads.

SEC

The SEC defines resources as:

Resources are quantities of oil and gas estimated to exist in naturally occurring accumulations. A portion of the resources may be estimated to be recoverable, and another portion may be considered to be unrecoverable. Resources include

---

26 SPE/WPC Petroleum Resources Management System; March 2007, p.26.

both discovered and undiscovered accumulations.[27]

Resources are differentiated from reserves as reserves are in known accumulations and are economically producible. Reserves also require a legal right to produce or the expectation of such a right. Resource quantities are not reportable under SEC guidelines.

China

The Petroleum Reserve Standard defines the Total Mineral Resource as the sum of the Geologic Reserve and the Perspective Reserve. [28]

Geologic Reserves is the total producible volume in the reservoir at original subsurface conditions. The different classes of geologic reserves require different amounts of data and reflect various phases of exploration and development. Simplified models are used in the exploration and development phrases especially to define water contacts in complex or multilayered reservoirs such as those encountered in a fluvial depositional system. Geologic reserves are divided into **In-Table Reserves**, which are reserves that are economic under current operating conditions and prices, and **Off-Table Reserves** which are not currently economic but could be with higher prices or improved technology.[29]

The PRC notes two types of resources, the Potential Resource and the Speculative Resource as a part of the overall, broader classification of Perspective Resource. Resource classifications are based on the maturity of the data acquisition and the amount of information available to quantify the accumulation.

The Perspective Resource in China is defined as undiscovered hydrocarbons, much as the AAPG definition of Prospective Reserves. They are evaluated using

27 SECURITIES AND EXCHANGE COMMISSION, TITLE 17, Code of Federal Regulations Parts 210, 211, 229, and 249, MODERNIZATION OF OIL AND GAS REPORTING, p. 7.

28 The National Standard of P.R.C., Petroleum Reserve Standard.

29 Id.

probabilistic methods based on geological and geophysical data.[30] The reserve volume is approximate. Perspective resources are divided into two categories, potential and speculative.

Potential Resource

The Potential Resource or "Perspective Resource by Trapped Method" is evaluated based on geological and geophysical data. All of the known traps in the area are evaluated and provide the basis of the parameters and volumes assigned.[31] It is the first step in the exploration process.

Speculative Resource

The Speculative Resource is based on aerial photography and analogy to the adjacent basin geology.[32] The resource is evaluated using geochemical data and information from test wells as well as analogy to nearby basins. It is based on a probabilistic analysis. This is the basis for the long term exploration program.

30 The National Standard of P.R.C., Petroleum Reserve Standard.

31 Id.

32 Id.

Reserves

Reserves are quantities of hydrocarbons which are commercially recoverable from known accumulations from a given date forward.[33] Reserves and the cumulative production combine into the ultimate recoverable volume.

SEC

The SEC initially wrote its reserve definitions in 1978 primarily for the financial markets and discussed only proved reserves. These definitions were written to provide a consistent reserves reporting basis for the financial community but, in many ways, they were not adequate for the planning purposes of exploration and production companies. In December 2008, the SEC revised its guidelines and reporting standards.

With the 2008 revision, the SEC defines reserves as "... the estimated remaining quantities of oil and gas and related substances anticipated being economically producible, as of a given date, by application of development projects to known accumulations."[34] They also add there must also "...exist, or there must be a reasonable expectation that there will exist, the legal right to produce or a revenue interest in the production of oil and gas, installed means of delivering oil and gas or related substances to market, and all permits and financing required to implement the project."[35] One difference between the SEC and the PRMS definition of reserves is the use of the term "economic producibility" rather than "commerciality" as used by the PRMS. The SEC's objection to the term commerciality is it implies the use of a rate of return which may vary from company to company.

33 Petroleum Resources Classification System and Definitions, Approved by the Board of Directors, Society of Petroleum Engineers (SPE) Inc., the Executive Board, World Petroleum Congresses (WPC), and the Executive Committee, American Association of Petroleum Geologists (AAPG), February 2000 "Reserves are defined as those quantities of petroleum which are anticipated to be commercially recovered from known accumulations from a given date forward."

34 Rule 4-10(a)(26) [17 CFR 210.4-10(a)(26)].

35 Security and Exchange Commission, Modernization of Oil and Gas Reporting, December 31, 2008, p.41.

The SEC also notes situations in which reserves should not be assigned which include: areas separated from the known reservoir by large, potentially sealing faults; and areas separated from the known reservoir by a non-productive area such as by a structurally low area, an area with no reservoir rock or an area with a poor test.[36] These accumulations are more properly referred to as resources.

With the revisions to the reserve standards, the SEC now allows reporting of probable and possible reserves, although they do not require the reporting of these volumes. Proved volumes include volumes of which there is a reasonable certainty of commercial recovery. Proved plus probable volume estimates are less certain but still more likely than not to be recovered.[37] Proved plus probable plus possible volume estimates might be achieved, but only under more favorable circumstances than are likely. [38]

SPE

In February 2000, the SPE/WPC and AAPG developed a set of broad reserve definitions covering the complete range of petroleum resources and reserves. The definitions begin with "total petroleum-initially-in-place, which is the quantity of hydrocarbons estimated to have originally existed in natural accumulations.[39] These are then divided into discovered and undiscovered categories. The

---

36 Note to Rule 4-10(a)(26) [17 CFR 210.4-10(a)(26)].

37 17 CFR 210.4-10(a)(18), "Probable reserves are those additional reserves that are less certain to be recovered than proved reserves but which, together with proved reserves, are as likely as not to be recovered."

38 17 CFR 210.4-10(a)(18), "Possible reserves are those additional reserves that are less certain to be recovered than probable reserves."

39 Petroleum Resources Classification System and Definitions, Approved by the Board of Directors, Society of Petroleum Engineers (SPE) Inc., the Executive Board, World Petroleum Congresses (WPC), and the Executive Committee, American Association of Petroleum Geologists (AAPG), February 2000 "Total Petroleum-initially-in-place is that quantity of petroleum which is estimated to exist originally in naturally occurring accumulations. Total Petroleum-initially-in-place is, therefore, that quantity of petroleum which is estimated, on a given date, to be contained in known accumulations, plus those quantities already produced therefrom, plus those estimated quantities in accumulations yet to be discovered. Total Petroleum-initially-in-place may be subdivided into Discovered Petroleum-initially-in-place and Undiscovered Petroleum-initially-in-place, with Discovered Petroleum-initially-in-place being limited to known accumulations."

discovered petroleum-initially-in-place includes what will be discussed in this paper as reserves, as well as contingent resources. [40]

The SPE/WPC uses a definition for proved reserves similar to that of the SEC, and also recognizes probable and possible reserves. These categories are less certain than proved and involve more uncertainty and risk than proved reserves, but are useful for oil and gas company planning purposes.

PRC

In 1994 the Ministry of Land and Resources of China revised the 1988 reserve definitions. China now uses a reserve hierarchy similar to the SPE/WPC. The PRC has refined its reserve definitions to more closely follow the SPE/WPC definitions and still keeping the basic features of the old system. Both the old and new reserve definitions include proved, probable and possible reserves.

General

Proved reserves are sometimes denoted as 1P. Probable and possible reserves are denoted as P2 and P3 respectively. More often the reserves are called 2P, which is proved plus probable or 3P, which is proved plus probable plus possible.

Risk as used here refers to the possibility that the reserves will or will not be there. Uncertainty refers to the lack of confidence as to the size or producibility of the accumulation. Many reservoir parameters cannot be measured directly and are inferred or estimated by analysis of various pieces of geological and engineering data. As a consequence, there is uncertainty in any reserve estimate.

---

40 Petroleum Resources Classification System and Definitions, Approved by the Board of Directors, Society of Petroleum Engineers (SPE) Inc., the Executive Board, World Petroleum Congresses (WPC), and the Executive Committee, American Association of Petroleum Geologists (AAPG), February 2000 "Discovered Petroleum-initially-in-place is that quantity of petroleum which is estimated, on a given date, to be contained in known accumulations, plus those quantities already produced therefrom. Discovered Petroleum-initially-in-place may be subdivided into Commercial and Sub-commercial categories, with the estimated potentially recoverable portion being classified as Reserves and Contingent Resources respectively ..."

Oil and Gas Reserve Guidelines

Different reserve classifications infer varying amounts of uncertainty concerning the reserves and for this reason should not be added together. Each should be considered separately.

Minerals Other than Oil and Gas

Mineral reserves and resources other than oil and gas are classified according to standards set out by the Council for Mining and Metallurgical Institutions (CMMI) as measured, indicated or inferred resources. Measured resources have been sampled and tested at locations which are spaced close enough to confirm geological continuity. This requires a high level of confidence and understanding of the geology.[41] Indicated resources are sampled at locations too far apart to confirm the geologic continuity with a high degree of certainty, but does allow for a reasonable level of certainty.[42] Inferred resources have a low level of confidence.[43] Oil and gas resources are Contingent and Prospective.

Mineral reserves are measured or indicated resources which have been shown to be able to justify exploitation under realistic economic and technical conditions.[44]

Mineral reserves are based on deterministic evaluations, while oil and gas reserves can be based on deterministic or probabilistic evaluations. Mineral reserves are classified as either proved or probable, while oil and gas reserves can be classified as proved, probable or possible.

41 Cronquist, Chapman, Estimation and Classification of Reserves of Crude oil, Natural Gas, and Condensate, 2001.

42 Id.

43 Id.

44 Id.

Unconventional Resources

Under its new guidelines, the SEC treats nontraditional or unconventional resources more like conventional resources for reporting purposes. Under the proposed rule changes for 2008, the SEC proposed using the term "continuous reservoirs"[45] Continuous reservoirs are defined as accumulations with diffuse boundaries, with hydrocarbons not necessarily accumulated above or bounded by hydrocarbon-water contacts."[46] Under the new guidelines, the SEC limits unconventional resources to "non-renewable resources which oil or gas can be extracted from".[47]

Unconventional resources included in this discussion are oil sands, tar sands, oil shale, fractured shale, and heavy oil. Also considered as unconventional resources are gas shale, tight gas sands coalbed methane, and gas hydrates. The oil sands are actually bitumen, while oil shale contains kerogen which can be converted into oil. The volumes of these hydrocarbons are large, and the improved technology is required to recover much of it. Unconventional resources are generally accessible only at high prices and production is more analogous to mining than conventional oil and gas production. Production is characterized by longer lead times than conventional oil and gas to peak production, and after a period of flat production a slow steady decline can be noted.

Oil shale is rich in organic content and when processed, yields kerogen which must then be further processed into oil. Most of the world's oil shale deposits can be found in the western United States. The oil shale deposits in the Piceance Basin of Colorado have been reported to yield upwards of 10 gallons per ton.

45 Security and Exchange Commission, Modernization of Oil and Gas Reporting, Conforming Version, June 26, 2008, p.41.

46 Reserves in Nontraditional Reservoirs: How Can We Account for Them, Lee, W.J. SPE-123384, October 2009, p.2.

47 Id.

One of the major hurdles of oil shale is the disposal of the waste material, some of which can be toxic and the large amount of water involved. Because of the difficulty in extracting the kerogen and disposing of the waste, it is unlikely there will be significant amounts of oil generated from oil shale. Deposits can also be found in Brazil, Zaire, China, Australia and the FSU. Although there are large resources available, there is not a significant amount of production from oil shale at present. Costs have been estimated to run about $11.50 per barrel for the deposits in Australia.

Bitumen, asphalt and tar are composed of oil, from which the light ends have been removed, and sand. They grade into heavy oils. The dividing line between these is arbitrary. The cutoff of 10 degrees gravity and 10 mpas is often used between heavy oil and bitumen, with the upper end of heavy oil at as much as 20 degrees gravity. Asphalt is bitumen with a gravity of about zero. There are large deposits of heavy oil and bitumen in Canada and Venezuela as well as Indonesia, China, Brazil, and the former Soviet Union. Unlike the oil shale, there is significant production from oil sands.

Bitumen and heavy oil can be strip mined if the deposits are shallow enough, generally less than 250'. It takes about two tons of oil sand to produce one barrel of oil. The oil sand is mined and made into slurry by mixing it with water. The sand is then separated from the oil using air and chemicals. If the deposits are deeper, steam floods are used to reduce the viscosity and make the oil able to flow. One of the newer methods involves horizontal wells with two laterals, one above the other, referred as the SAGD method or Steam Assisted Gravity Drainage. The upper lateral is used for steam injection and the lower is used to produce the oil by gravity drainage. Some estimates indicate about 20% of the Canadian oil sand can be developed by surface mines. One of the major concerns of producing the deeper sands is large amount of natural gas needed to heat the oil.

The U.S. Geological Survey has begun to recognize bitumen as reserves. The SEC previously said that oil "manufactured" from gilsonite, oil shale or coal is a mining activity and should be reported under Industry Guide 7 as such rather than oil and gas reserves. With the December 31, 2008 revisions, the SEC allows reporting of oil and gas from oil and gas shales and bitumen. The SEC defines bitumen as follows: "Bitumen, sometimes referred to as natural bitumen, is petroleum in a solid or semi-solid state in natural deposits with a viscosity greater than 10,000 centipoise measured at original temperature in the deposit and atmospheric pressure, on a gas free basis. In its natural state it usually contains sulfur, metals, and other non-hydrocarbons."[48]

Canada has published guidelines for reporting Mineable Oil Sand Reserves. The reports must be prepared by a qualified person who is defined to be an engineer or geoscientist with at least five years experience in mineable oil sands and is a member of a self-regulating organization. The person should be able to show they have sought the appropriate expertise from others if they do not posses it themselves. The CIM standards require the completion of a Preliminary Feasibility Study as a minimum to move oil sand resources to reserves. The study should show the viability of the method and mine configuration and an effective method of processing the oil sand. A financial analysis must be included. Exploration information is reported in the form of tonnage and grade and the qualified person should state if they are conceptual or order of magnitude.

The CIM defines a minable oil sands resource as an occurrence of bitumen sand of such quantity and grade for reasonable prospects of economic production. Mineable oil sands reserves are economically recoverable as demonstrated by a preliminary feasibility study.

Although not previously allowed, the SEC, with the December 31,2008 revision, now allows reporting of unconventional resources such as bitumen from oil sands and oil or gas from coal and shale.[49] The SEC has revised the definitions to "...include the extraction of saleable hydrocarbons, in the solid, liquid, or gaseous state, from oil sands, shale, coalbeds, or other nonrenewable natural

48 Security and Exchange Commission, Modernization of Oil and Gas Reporting, December 31, 2008, p.133.

49 Rule 4-10(a)(1)(ii)(D) [17 CFR 210.4(a)(1)(ii)(D)].

resources which are intended to be upgraded into synthetic oil or gas, and activities undertaken with a view to such extraction."[50] In a footnote, the SEC also says a product is "salable" if it is a state in which it can be sold. There does not have to be a market for product to be "salable".[51]

The SEC does recognize a potential problem with reporting oil and gas reserves from coal and oil shale in that both can be used as fuel as is. The SEC, therefore, requires a company to book oil and gas reserves for coal or oil shale intended to be converted to oil or gas and not to book oil and gas reserves for coal or oil shale not intended to be converted to oil or gas.[52] The SEC also requires a separate disclosure for traditional and for synthetic oil or gas, as they believe this distinction is important to investors.[53]

The SEC has treated certain unconventional resources as reserves for some time. For instance, coalbed methane gas could be considered reserves under the old SEC definitions. Coalbed methane is also referred to as natural gas from coal in Canada and coal seam gas in Australia. Gas from shale was included under the old rules if produced in place. Under the new rules, gas reserves from coal or shale can be reported whether they are natural or synthetic. Oil and gas can now be treated as proved reserves whether created in place or by a processing method if they meet the other criteria for reserves.[54] These criteria include economic production, and assurance that permits, including governmental permits, have been obtained.

---

50 Security and Exchange Commission, Modernization of Oil and Gas Reporting, December 31, 2008, p.22.

51 Security and Exchange Commission, Modernization of Oil and Gas Reporting, December 31, 2008, p.22. "A hydrocarbon product is saleable if it is in a state in which it can be sold even if there is no ready market for that hydrocarbon product in the geographic location of the project. The absence of a market does not preclude the activity from being considered an oil and gas producing activity. However, in order to claim reserves for that hydrocarbon product from a particular location, there must be a market, or a reasonable expectation of a market, for that product."

52 Security and Exchange Commission, Modernization of Oil and Gas Reporting, December 31, 2008, p.23.

53 Security and Exchange Commission, Modernization of Oil and Gas Reporting, December 31, 2008, p.24.

54 Reserves in Nontraditional Reservoirs: How Can We Account for Them, Lee, W. J. SPE Economics and Management, October 2009, p.12.

Coalbed methane or CBM is generated in coal and remains in the coal due to absorption. There are two types of gas normally referred to with CBM – free gas or gas in the fractures or cleats and absorbed gas which is liberated from the coal as the pressure is reduced. The coal is both the source and the reservoir. Coal deposits are typically terrestrial and include low energy deltaic environments. The gas is usually sweet with few impurities.

Shale oil and shale gas are produced from organic rich mudrocks. These rocks are the source and reservoir. Because of the low matrix permeability, they require fracturing to produce at economic rates. Typically shale reservoirs are marine and low energy environments. These reservoirs may have significant volumes of impurities such as CO_2.

Shale plays usually cover a large area. They are low porosity and permeability reservoirs which require fracturing for economic production. The hydrocarbons are generated within the reservoir.

Since 2011, the SEC has taken a special interest in shale gas reserves. They are looking into decline curve and type curves, profitability, the five year PUD rule and hydraulic fracturing.[55]

In 2011 claims were made by certain people that shale gas companies were overestimating reserves and underestimating costs. The SEC investigated a number of companies. Subpoenas were send out requesting information on reserve calculations including what formulas were used, what assumptions were made, decline curves and type curves, and how the actual production compared to the forecast production.[56] They also looked at the development of proved undeveloped reserves and if they were being developed within the five year time frame, reliable technology being used, terminal decline rates, well life,

55 Securities Litigation and Enforcement Risks for Shale Operators; G Pecht and P. Stokes, Fulbright & Jaworski L.L.P.

56 Id.

and significant changes reported in proved reserves.[57] Companies should have a reasonable basis for the reserve estimate and be able to explain and justify the assumptions made.

Under the 2008 rule changes, proved undeveloped reserves could be assigned more than one location from the producing well. "Reliable technologies" could be used to establish these reserves and not just offset production. As a result, many companies having shale gas reserves were able to significantly increase their reserve estimates by adding undeveloped locations. The additional locations were still subject to the five year rule – the companies must have a development plan showing with reasonable certainty these reserves will be developed within five years.

Gas prices have fallen in recent years and the question now is how much the companies are going to have to write off due to economics. Lower prices have produced lower profit margins and reduced drilling for gas.

---

57 Securities Litigation and Enforcement Risks for Shale Operators; G Pecht and P. Stokes, Fulbright & Jaworski L.L.P.

Problems with Reserve Classification

The complexity of reserve classification requires that each case be considered a unique occurrence and evaluated as such. It is difficult to write rules to cover each possible situation. Reserve estimates can be imprecise due to parameter uncertainties and the limited nature of the data upon which the estimate is based.[58]

The prior SEC rules were adopted in 1978, and since then, there has been much advancement in technology. The December 31, 2008 revisions allow for the use of some of the new technology available to the oil and gas industry. There have been improvements in the acquisition and processing of seismic data that provide better reservoir definition than possible in 1978. There has been the advent of MDTs to better define the reservoir's ability to produce and its limits. The increased availability of personal computers has made probabilistic methods easier to use and more widespread. The SEC now allows the use of "reliable technology" which it defines as "Reliable technology is a grouping of one or more technologies (including computational methods) that have been field tested and has been demonstrated to provide reasonably certain results with consistency and repeatability in the formation being evaluated or in an analogous formation."[59]

The recent write-downs by major oil and gas companies give some clues to the complexity of the reserve reporting process. Until all of the oil or gas in a reservoir is produced, the reserve estimate will likely change from year to year depending on: 1) the acquisition of new data supporting the reservoir size and characteristics, 2) operating conditions and 3) economics. Since much of

58 Auditing Standards for Reserves, SPE, June 2001. "The reliability of Reserve Information is considerably affected by several factors. Initially, it should be noted that Reserve Information is imprecise due to the inherent uncertainties in, and the limited nature of, the database upon which the estimating and auditing of Reserve Information is predicated. Moreover, the methods and data used in estimating Reserve Information are often necessarily indirect or analogical in character rather than direct or deductive.

59 Security and Exchange Commission, Modernization of Oil and Gas Reporting, December 31, 2008, p. 141.

the reserve characterization is subjective, or data may be lacking, there can be significant differences in reserve estimates, depending on the experience of the evaluator and the quality of the data available to him. Changes in foreign contracts or the makeup of foreign governments can cause reserve category changes. Uncertainty regarding production economics or governmental regulations can also make a reserve classification change necessary.

In discussing reasonable certainty, the SEC says it means a high degree of confidence the reserves will be recovered, and says a high degree of certainty "... exists if the quantity is much more likely to be achieved than not, and, as changes due to increased availability of geoscience (geological, geophysical, and geochemical), engineering, and economic data are made to estimated ultimate recovery (EUR) with time, reasonably certain EUR is much more likely to increase or remain constant than to decrease."[60]

60 Security and Exchange Commission, Modernization of Oil and Gas Reporting, December 31, 2008, p. 140-141.

Proved Reserves

The SEC defined proved reserves in Rule 4-10(a) of Regulation S-X of the Securities and Exchange Act of 1934 (see Appendix B) as those quantities of hydrocarbons that can be shown with a reasonable certainty to be economically recoverable under current economic and operating conditions from known reservoirs in future years.[61] With the December 31, 2008 guidelines, the SEC has revised its definition to "Reserves are estimated remaining quantities of oil and gas and related substances anticipated to be economically producible, as of a given date, by application of development projects to known accumulations. In addition, there must exist, or there must be a reasonable expectation that there will exist, the legal right to produce or a revenue interest in the production, installed means of delivering oil and gas or related substances to market, and all permits and financing required to implement the project."[62]

Rule 4-10 of SEC Regulation S-X was the primary source of information on SEC reserve definitions. Since this regulation was written in 1978, however, various staff accounting bulletins have been written to expand the industry's understanding of the definitions and reporting requirements and the definitions and guidelines revised December 31,2008.

Similar to the SEC, the SPE/WPC defined proved reserves as petroleum quantities that can be commercially recovered from a given date forward from known reservoirs, with reasonable certainty, and under current conditions[63]. If

61 Rule 4-10(a-2) Regulation S-X, Securities and Exchange Act of 1934. "Proved oil and gas reserves are the estimated quantities of crude oil, natural gas, and natural gas liquids which geological and engineering data demonstrate with reasonable certainty to be recoverable in future years from known reservoirs under existing economic and operating conditions, i.e., prices and costs as of the date the estimate is made. Prices include consideration of changes in existing prices provided only by contractual arrangements, but not on escalations based upon future conditions."

62 Security and Exchange Commission, Modernization of Oil and Gas Reporting, December 31, 2008, p. 141.

63 SPE/WPC Reserve Definitions "Proved reserves are those quantities of petroleum which, by analysis of geological and engineering data, can be estimated with reasonable certainty to be commercially recoverable, from a given date forward, from known reservoirs and under current economic conditions, operating methods, and government regulations. Proved reserves can be categorized as developed or undeveloped."

the probabilistic method is used, the SPE/WPC suggests a 90% probability the amount recovered will equal or exceed the reserve estimate. If deterministic methods are used, the SPE and SEC require a high degree of confidence the hydrocarbons will be recovered.

The PRMS and SEC differ in their definitions as the PRMS uses the term commerciality and the SEC uses the term economically producible. The SEC felt using the term commerciality implied the use of a rate of return before a company would commit to a project. This would add one more variable into the definition, since the rate of return required by various companies is different, and thus reduce the ability of investors to evaluate companies.

Under the December 31, 2008 revisions, the SEC defines proved reserves as: "Proved oil and gas reserves are those quantities of oil and gas, which, by analysis of geoscience and engineering data, can be estimated with reasonable certainty to be economically producible—from a given date forward, from known reservoirs, and under existing economic conditions, operating methods, and government regulations—prior to the time at which contracts providing the right to operate expire, unless evidence indicates that renewal is reasonably certain, regardless of whether deterministic or probabilistic methods are used for the estimation. The project to extract the hydrocarbons must have commenced or the operator must be reasonably certain that it will commence the project within a reasonable time.

(i) The area of the reservoir considered as proved includes:

(A) The area identified by drilling and limited by fluid contacts, if any, and

(B) Adjacent undrilled portions of the reservoir that can, with reasonable certainty, be judged to be continuous with it and to contain economically producible oil or gas on the basis of available geoscience and engineering data.

(ii) In the absence of data on fluid contacts, proved quantities in a reservoir are limited by the lowest known hydrocarbons (LKH) as seen in a well penetration unless geoscience, engineering, or performance data and reliable technology establishes a lower contact with reasonable certainty.

(iii) Where direct observation from well penetrations has defined a highest known oil (HKO) elevation and the potential exists for an associated gas cap, proved oil reserves may be assigned in the structurally higher portions of the reservoir only if geoscience, engineering, or performance data and reliable technology establish the higher contact with reasonable certainty.

(iv) Reserves which can be produced economically through application of improved recovery techniques (including, but not limited to, fluid injection) are included in the proved classification when:

(A) Successful testing by a pilot project in an area of the reservoir with properties no more favorable than in the reservoir as a whole, the operation of an installed program in the reservoir or an analogous reservoir, or other evidence using reliable technology establishes the reasonable certainty of the engineering analysis on which the project or program was based;

(B) The project has been approved for development by all necessary parties and entities, including governmental entities.

(v) Existing economic conditions include prices and costs at which economic producibility from a reservoir is to be determined. The price shall be the average price during the 12-month period prior to the ending date of the period covered by the report, determined as an unweighted arithmetic average of the first-day-of-the-month price for each month within such period, unless prices are defined by contractual arrangements, excluding escalations based upon future conditions."[64]

"If deterministic methods are used, reasonable certainty means a high degree

64 Security and Exchange Commission, Modernization of Oil and Gas Reporting, December 31, 2008, p. 139-140.

of confidence that the quantities will be recovered. If probabilistic methods are used, there should be at least a 90% probability that the quantities actually recovered will equal or exceed the estimate."[65]

Proved reserves are placed in one of two major categories; either proved developed or proved undeveloped by both the SEC and the SPE/WPC. The SPE/WPC divides proved developed reserves into producing and non-producing, with non-producing including shut-in and behind pipe reserves. The SEC does not require the reporting of other than developed or undeveloped reserves.

The SPE/WPC has expanded its reserve classifications to include economic modifiers including economic (reserves), marginal economic (contingent resources) and sub-marginal economic (contingent resources).[66]

The SEC definitions deal mainly with deterministic reserve estimates. The SEC recognizes probabilistic reserve estimates only to the extent they do not go beyond their definitions, for example, include volumes below a lowest know hydrocarbon. Aggregation is not permitted beyond the field or property level by the PRMS and similarly the SEC says "Regardless of whether the reserves were determined using deterministic or probabilistic methods, the reported reserves should be simple arithmetic sums of all estimates at the well, reservoir, property, field, or project level within each reserves category."[67] The SPE/WPC recognizes reserves based on either the deterministic or probabilistic method.[68]

China divides its Total Oil and Gas Resources into Discovered and Undiscovered Reserves. Discovered Reserves or Geologic Reserves include the original oil and gas in place and are classified using a three class system. The classification of reserves is based on the maturity of the project and knowledge of the reservoirs.

65 Modernization of Oil and Gas Reporting CFR 210.4-10 (24).

66 SPE/WPC Petroleum Resources Management System; March 2007.

67 Security and Exchange Commission, Modernization of Oil and Gas Reporting, December 31, 2008, p. 60.

68 SPE/WPC Reserve Definitions "The method of estimation is called deterministic if a single best estimate of reserves is made based on known geological, engineering, and economic data. The method of estimation is called probabilistic when the known geological, engineering, and economic data are used to generate a range of estimates and their associated probabilities."

Inferred reserves are reserves in the discovery or early exploration stage and have a relatively low level of confidence. Indicated reserves have had a wildcat well drilled and tested at commercial rates. Indicated reserves should have a moderate degree of confidence and a relative error of no more than 50%.[69] Measured reserves are those at the final stage of exploration and development and are estimated with a high level of confidence. They should have a relative error of no more than 20%.[70]

The PRC's Recoverable Reserve volumes are the recoverable hydrocarbons from Geologic Reserves. They are divided into seven classifications depending on the degree of geologic certainty and economic viability.

Recoverable Reserves are classified as: [71]

Proved Technically Estimated Ultimate Recoveries which are technically estimates ultimate recoveries which have the current technology to be produced, have a development plan and are economic or sub-economic to develop based on current prices and costs.

Proved Economic Initially Recoverable Reserves are reserves which have the current technology to be produced, have a development plan, are economic under current prices. The reservoir boundaries are defined by drilling or pressure

---

69 GAO Ruiqi, LU Minggang, ZHA Quanheng, XIAO Deming, HU Yundong, China Petroleum Resources/Reserves Classification. "5.2.2.2 ***Indicated Geological Reserves (IDGR):*** Indicated geological reserves are estimated with a moderate level of confidence and relative error not more than ±50 when oil and/or gas economic flow is obtained from prospecting well at general exploration phase. The estimation of indicated geological reserves should preliminarily ascertaining structure configuration, formation continuity, oil and gas distribution, reservoir type, fluid properties and productivities, etc. The geological confidence degree is moderate, which can be as evidence for drilling reservoir appraisal wells, making conceptual design or development plan."

70 GAO Ruiqi, LU Minggang, ZHA Quanheng, XIAO Deming, HU Yundong, China Petroleum Resources/Reserves Classification. "5.2.2.1 ***Measured Geological Reserves (MEGR):*** Measured geological reserves are estimated with a high level of confidence and relative error not more than ±20 after the reservoirs have been proved economically recoverable by appraisal drilling at appraisal phase. The estimation of measured geological reserves should identifying the reservoir type, pore morphology, drive mechanism, fluid properties, distributions and productivities, etc. Fluid contacts or the lowest known hydrocarbon should be determined by drilling, logging and test data or reliable pressure data. The reasonable well spacing or primary development well pattern should be used in the delineation of measured limits. All parameters in the volumetric approach should have a high degree of certainty."

71 GAO Ruiqi, LU Minggang, ZHA Quanheng, XIAO Deming, HU Yundong, China Petroleum Resources/Reserves Classification.

data or the lowest known hydrocarbon is used. They have a sales contract and a pipeline. There should be at least an 80% probability the actual volumes recovered will equal or exceed the estimated volumes.

Proved Sub-economic Initially Recoverable Reserves are the difference between the Proved Technically Estimated Ultimate Recoveries and the Proved Economic Initially Recoverable Reserves.

Probable Technically Estimated Ultimate Reserves are technically recoverable, but are not economic.

Probable Economic Initially Recoverable Reserves are economic to develop and there is at least a 50% probability the volumes recovered will equal or exceed the estimated Economic Initially Recoverable Reserves.

Probable Sub-economic Initially Recoverable Reserves are the difference between the Probable Technically Estimated Ultimate Recoveries and the Probable Economic Initially Recoverable Reserves.

Possible Technically Estimated Ultimate Reserves are quantities which have at least a 10% probability of being recovered in the future and exceed the Economic Initially Recoverable Reserves.

The PRC defines proved reserves as estimated quantities that production or tests show is economic or by analogy is shown to be economic in other wells; the reservoir boundary has been determined based on drilling or pressure data; has a development plan; implemented operation or pilot or analogy to a similar field; evaluation based on the cost and price on the evaluation date or based on contract is economic; and future production should be equal to or greater than 80% of the reserve estimate.

The PRC indicates proved reserves can be assigned when the appraisal drilling is nearly complete. They form the basis of the plan of development. The reservoir parameters and limits can be reliably estimated.

Proved Developed

The SEC defines developed reserves as reserves able to be produced economically by existing wells using existing equipment and operating conditions. The December 31, 2008 revisions define reserves as developed, if "... the cost of any required equipment is relatively minor compared to the cost of a new well."[72] This is similar to the PRMS definition. Improved recovery projects are considered developed only after a successful pilot project showing an actual production increase from the program.[73] Reserves are considered developed if the majority of the expenditures to develop the reserves have been spent. The SEC does not subdivide proved developed into producing and non-producing.

There are factors which can cause non producing reserves to be placed in a category less than proved, although otherwise the reserves would be proved.[74] These include the lack of a market, waiting on stimulation, not enough estimated reserves for a pipeline or way to get the reserves to market or mechanical problems. Proved reserves should not be assigned until there is sufficient information to confirm the commerciality of the project and thus the existence of reserves.

The SEC gives examples of different types of wells they would consider proved developed.[75] These include producing wells, or wells that require a minimum

---

72 See Rule 4-10(a)(6) [17 CFR 210.4-10(a)(6)].

73 Rule 4-10(a-3) Regulation S-X, Securities and Exchange Act of 1934. "Proved developed oil and gas reserves are reserves that can be expected to be recovered through existing wells with existing equipment and operating methods. Additional oil and gas expected to be obtained through the application of fluid injection or other improved recovery techniques for supplementing the natural forces and mechanisms of primary recovery should be included as 'proved developed reserves' only after testing by a pilot project or after the operation of an installed program has confirmed through production response that increased recovery will be achieved."

74 Reserves Definitions Committee Society of Petroleum Evaluation Engineers, "Guidelines for Application of Petroleum Reserves Definitions", p.15.

75 SEC Division of Corporation Finance: Frequently Requested Accounting and Financial Reporting Interpretations and Guidance, March 31, 2001 "Currently producing wells and wells awaiting minor sales connection expenditure, recompletion, additional perforations or bore hole stimulation treatment would be examples of properties with proved developed reserves since the majority of the expenditures to develop the reserves has already been spent."

expenditure to be able to produce, such as needing a minor sales connection, additional perforations, or stimulation.

The SEC indicates that proved developed reserves can be assigned to enhanced recovery operations if there is a successful pilot project or the project is fully installed and operational and a response has been seen.[76] If the reserves are based on a pilot project, then only the portion of the reservoir affected by the pilot can be assigned proved developed reserves. If the project is installed and operational, developed reserves can be assigned the part currently seeing a response. Proved developed reserves can be assigned to an improved recovery project only to the extent they are supported by actual production. If the actual production is less than the anticipated and projected performance, only the actual can be classified as proved developed. The SEC notes it is important for facilities to be installed for the improved recovery project and a production increase seen before the reserves are considered proved developed.

The SPE/WPC defines proved developed reserves as reserves expected to be recovered from existing wells. They require facilities to be in place for improved recovery projects.[77] The requirement for a pilot or the successful initiation of a project is still required before the reserves are considered proved.

The SPE/WPC also subdivides proved developed reserves into producing[78] and

---

76 SEC Division of Corporation Finance: Frequently Requested Accounting and Financial Reporting Interpretations and Guidance, March 31, 2001 "Proved developed reserves from improved recovery techniques can be assigned after either the operation of an installed pilot program shows a positive production response to the technique or the project is fully installed and operational and has shown the production response anticipated by earlier feasibility studies. In the case with a pilot, proved developed reserves can be assigned only to that volume attributable to the pilot's influence. In the case of the fully installed project, response must be seen from the full project before all the proved developed reserves estimated can be assigned. If a project is not following original forecasts, proved developed reserves can only be assigned to the extent actually supported by the current performance. An important point here is that attribution of incremental proved developed reserves from the application of improved recovery techniques requires the installation of facilities and a production increase."

77 SPE/WPC Reserve Definitions "Developed reserves are expected to be recovered from existing wells including reserves behind pipe. Improved recovery reserves are considered developed only after the necessary equipment has been installed, or when the costs to do so are relatively minor. Developed reserves may be sub-categorized as producing or non-producing."

78 SPE/WPC Reserve Definitions "Reserves subcategorized as producing are expected to be recovered from completion intervals which are open and producing at the time of the estimate. Improved recovery reserves are considered producing only after the improved recovery project is in operation."

non-producing.[79] Producing reserves are attributed to zones that are open and producing at the time of the evaluation. Non-producing reserves are further subdivided into shut-in and behind pipe. Shut-in reserves are allocated to zones perforated and open, but are not producing at the time of the evaluation. Behind-pipe reserves are reserves available from existing wells in zones which still require completion.

The Petroleum Reserve Standard of China defines proved developed reserves (I or A) as reserves identified as available under current technical and economic conditions, having a development plan in place, all production wells completed, all facilities in place, and the reserves are on production.[80] These reserves can be increased after facilities for improved recovery are installed. These reserves are used as the basis for development analysis and management.

79 SPE/WPC Reserve Definitions "Non-producing: Reserves subcategorized as non-producing include shut-in and behind-pipe reserves. Shut-in reserves are expected to be recovered from (1) completion intervals which are open at the time of the estimate but which have not started producing, (2) wells which were shut-in for market conditions or pipeline connections, or (3) wells not capable of production for mechanical reasons. Behind-pipe reserves are expected to be recovered from zones in existing wells, which will require additional completion work or future recompletion prior to the start of production."

80 The National Standard of P.R.C., Petroleum Reserve Standard.

Proved Undeveloped

Proved undeveloped reserves are assigned to wells which have not been drilled or in situations where there is a major amount of money still to be spent to get the wells on production or to have facilities installed. Both the SEC[81] and SPE/WPC[82] have similar definitions.

One of the most significant aspects of the revision for undeveloped locations is the replacement of the certainty test for areas beyond one offset drilling location with the reasonable certainty test.

SEC

General

Proved undeveloped reserves are perceived to have more risk than proved developed reserves. It is likely the SEC will review these in more detail than in the past in light of recent reserve write-downs and the mandates of Sarbanes-Oxley.

Definition

The SEC defines undeveloped oil and gas reserves as "Undeveloped oil and gas reserves are reserves of any category that are expected to be recovered from

---

81 Rule 4-10(a-4) Regulation S-X, Securities and Exchange Act of 1934 "Proved undeveloped oil and gas reserves are reserves that are expected to be recovered from new wells on undrilled acreage, or from existing wells where a relatively major expenditure is required for recompletion. Reserves on undrilled acreage shall be limited to those drilling units offsetting productive units that are reasonably certain of production when drilled. Proved reserves for other undrilled units can be claimed only where it can be demonstrated with certainty that there is continuity of production from the existing productive formation. Under no circumstances should estimates, for proved undeveloped reserves be attributable to any acreage for which an application of fluid injection or other improved recovery technique is contemplated, unless such techniques have been proved effective by actual tests in the area and in the same reservoir."

82 SPE/WPC Reserve Definitions "Undeveloped Reserves: Undeveloped reserves are expected to be recovered: (1) from new wells on undrilled acreage, (2) from deepening existing wells to a different reservoir, or (3) where a relatively large expenditure is required to (a) recomplete an existing well or (b) install production or transportation facilities for primary or improved recovery projects."

new wells on undrilled acreage, or from existing wells where a relatively major expenditure is required for recompletion." [83]

(i) Reserves on undrilled acreage shall be limited to those directly offsetting development spacing areas that are reasonably certain of production when drilled, unless evidence using reliable technology exists that establishes reasonable certainty of economic producibility at greater distances.[84]

(ii) Undrilled locations can be classified as having undeveloped reserves only if a development plan has been adopted indicating that they are scheduled to be drilled within five years, unless the specific circumstances, justify a longer time.

(iii) Under no circumstances shall estimates for undeveloped reserves be attributable to any acreage for which an application of fluid injection or other improved recovery technique is contemplated, unless such techniques have been proved effective by actual projects in the same reservoir or an analogous reservoir...or by other evidence using reliable technology establishing reasonable certainty.[85]

Reliable technology is defined by the SEC to mean "...a grouping of one or more technologies (including computational methods) that has been field tested and has been demonstrated to provide reasonably certain results with consistency and repeatability in the formation being evaluated or in an analogous formation."[86]

Offsets

The SEC will allow proved reserves for legal locations offsetting commercial

---

83 **Federal Register,** Vol. 74, No. 9, January 14, 2009, Rules and Regulations, p. 2192.

84 **Id.**

85 Security and Exchange Commission, Modernization of Oil and Gas Reporting, December 31, 2008, p. 142-143.

86 Security and Exchange Commission, Modernization of Oil and Gas Reporting, December 31, 2008, p. 141.

wells.[87] There must conclusive data to support the reserves and enough acreage for a legal location. Proved reserves cannot be assigned below the lowest known hydrocarbon in the reservoir. Up to eight proved locations can be set up as direct offsets to a commercial well, if the reservoir is large enough.

In the December 31, 2008 revision, the SEC has replaced the term "drilling unit" with the term "development spacing area" when talking about offset locations.

Under the old guidelines, the SEC would allow only direct offsets as proved undeveloped unless "certainty" of reservoir continuity could be demonstrated.[88] The requirement was thus more stringent than the general definition of proved reserves as the qualifier of "reasonable" was not included for undeveloped.[89] Under the new guidelines of December 31, 2008, the SEC does not limit proved locations to direct offsets, if the company can establish "...with reasonable certainty that these reserves are economically producible."[90] The SEC has added the word "reasonable" to the certainty requirement, so now "reasonable certainty" is required and not "certainty" for offsets more than one location from a productive well.

The requirement of certainty for reservoir continuity could not always be

---

87 SEC Division of Corporation Finance: Frequently Requested Accounting and Financial Reporting Interpretations and Guidance, March 31, 2001 "In order to attribute proved reserves to legal locations adjacent to such a well (i.e. offsets), there must be conclusive, unambiguous technical data which supports reasonable certainty of production of such volumes and sufficient legal acreage to economically justify the development without going below the shallower of the fluid contact or the LKH. In the absence of a fluid contact, no offsetting reservoir volume below the LKH from a well penetration shall be classified as proved."

88 SEC Division of Corporation Finance: Frequently Requested Accounting and Financial Reporting Interpretations and Guidance, March 31, 2001 "Proved reserves for other undrilled units can be claimed only where it can be demonstrated with certainty that there is continuity of production from the existing productive formation." (Emphasis added)

89 SEC Division of Corporation Finance: Frequently Requested Accounting and Financial Reporting Interpretations and Guidance, March 31, 2001 "(f) Proved undeveloped oil and gas reserves are reserves that are expected to be recovered from new wells on undrilled acreage, or from existing wells where a relatively major expenditure is required for recompletion. Reserves on undrilled acreage shall be limited to those drilling units offsetting productive units that are reasonably certain of production when drilled. Proved reserves for other undrilled units can be claimed only where it can be demonstrated with certainty that there is continuity of production from the existing productive formation. Under no circumstances should estimates of proved undeveloped reserves be attributable to any acreage for which an application of fluid injection or other improved recovery technique is contemplated, unless such techniques have been proved effective by actual tests in the area and in the same reservoir." (Emphasis added)

90 See Rule 4-10(a)(24)(ii) [17 CFR 210.4-10(a)(24)(ii)].

satisfied by geology, seismic or drilling success ratios. Evidence of pressure communication was needed to convince the SEC of reservoir continuity.[91] The SEC requirement of certainty of reservoir continuity was a difficult test to meet. If a company wanted to book reserves more than one offset away, they needed to be prepared to present a very complete, detailed, and documented case to the SEC. Under the new guidelines, showing a reasonable certainty of reservoir continuity should be a less daunting task.

The SEC staff points out that proved undeveloped reserves cannot be considered more than one location from a commercial well if there is only one well in the reservoir.[92] Under the old guidelines, if there are two or more wells in the reservoir, then locations between the wells can be set up as proved undeveloped if there is "certainty" of reservoir continuity, even though they are more than one legal offset from the productive wells.[93] Under the December 31, 2008 revisions,

91 SEC Division of Corporation Finance: Frequently Requested Accounting and Financial Reporting Interpretations and Guidance, March 31, 2001 "(f)...Generally, proved undeveloped reserves can be claimed only for legal and technically justified drainage areas offsetting an existing productive well (but structurally no lower than LKH). If there are at least two wells in the same reservoir which are separated by more than one legal location and which show communication (reservoir continuity), proved undeveloped reserves could be claimed between the two wells, even though the location in question might be more than an offset well location away from any of the wells. In this illustration, seismic data could be used to help support this claim by showing reservoir continuity between the wells, but the required data would be the conclusive evidence of communication from production or pressure tests. The SEC staff emphasizes that proved reserves cannot be claimed more than one offset location away from a productive well if there are no other wells in the reservoir, even though seismic data may exist. The use of high-quality, well calibrated seismic data can improve reservoir description for performing volumetrics (e.g. fluid contacts). However, seismic data is not an indicator of continuity of production and, therefore, can not be the sole indicator of additional proved reserves beyond the legal and technically justified drainage areas of wells that were drilled. Continuity of production would have to be demonstrated by something other than seismic data."

92 SEC Division of Corporation Finance: Frequently Requested Accounting and Financial Reporting Interpretations and Guidance, March 31, 2001 "If there are at least two wells in the same reservoir which are separated by more than one legal location and which show communication (reservoir continuity), proved undeveloped reserves could be claimed between the two wells, even though the location in question might be more than an offset well location away from any of the wells. In this illustration, seismic data could be used to help support this claim by showing reservoir continuity between the wells, but the required data would be the conclusive evidence of communication from production or pressure tests. The SEC staff emphasizes that proved reserves cannot be claimed more than one offset location away from a productive well if there are no other wells in the reservoir, even though seismic data may exist. The use of high-quality, well calibrated seismic data can improve reservoir description for performing volumetrics (e.g. fluid contacts). However, seismic data is not an indicator of continuity of production and, therefore, can not be the sole indicator of additional proved reserves beyond the legal and technically justified drainage areas of wells that were drilled. Continuity of production would have to be demonstrated by something other than seismic data."

93 Id.

the requirement is "reasonable certainty" rather than certainty.[94] The SEC points out, that although seismic data can be used to help support the requirement of reservoir continuity, there must be conclusive evidence of communication based on either production or pressure tests. Seismic data can be used to help define the reservoir limits, but it is not considered an indicator of reservoir continuity by the SEC.

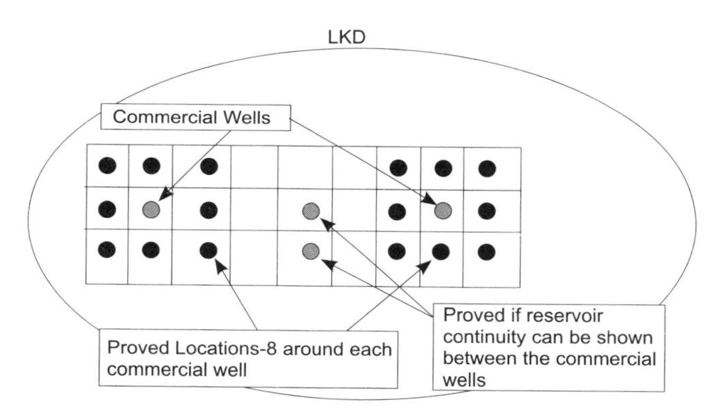

SEC Proved Locations

In coalbed methane reservoirs the same rules apply as for conventional reservoirs. The fact that the coal seam is continuous may not be sufficient in itself to establish the reasonable certainty requirement for reservoir continuity. The presence of coal by itself is not proof of continuity of permeability. In general, the rule of one location offset applies without reliable data to indicate continuity of production.

Horizontal wells have typically been treated differently than vertical wells by the SEC. The SEC has said that you should limit proved undeveloped reserves to two parallel offsets to the productive horizontal well. The SEC now will allow

94 See Rule 4-10(a)(24)(ii) [17 CFR 210.4-10(a)(24)(ii)].

proved undeveloped reserves offsetting the toe of a horizontal well.[95]

Undrilled Fault Blocks

The SEC's new guidelines do not allow assigning reserves to fault blocks which have not been penetrated, or separated from production by a non-productive reservoir.[96] The PRMS may allow these volumes to be classified as reserves if the evidence is compelling.

Reservoir Continuity

There are a number of ways to demonstrate reservoir continuity in a reservoir. Perhaps one of the best is to show the areas are in pressure communication using pressure data in different wells. Also reduced pressures in an area without production would be indicative of being in a common reservoir.

One can use log and seismic data to show reservoir continuity. Using subsurface correlations and seismic reflectors, sand continuity can be established and tied to a depositional model. If the seismic is not consistent and reliable, then it may not be accepted. If the reservoirs are laminated or discontinuous or variable in thickness or quality, then using logs and seismic to show continuity may not be accepted. An important factor to consider when using log and seismic data is if the data is not adequate in quality or quantity, the conclusion of reservoir continuity may not be accepted.

Commitment to Drill

Under the new SEC guidelines, undrilled locations can be classified as having

95 Compliance and Disclosure Interpretations, October 26, 2009 Question 131.01.

96 **Federal Register,** Vol. 74, No. 9, January 14, 2009, Rules and Regulations, p. 2192. "Reserves should not be assigned to adjacent reservoirs isolated by major, potentially sealing, faults until those reservoirs are penetrated and evaluated as economically producible. Reserves should not be assigned to areas that are clearly separated from a known accumulation by a non-productive reservoir (i.e., absence of reservoir, structurally low reservoir, or negative test results). Such areas may contain prospective resources (i.e., potentially recoverable resources from undiscovered accumulations)."

proved undeveloped reserves only if a development plan has been adopted.

The SEC requires a commitment by producers showing they will develop the infrastructure necessary to produce the undeveloped reserves.[97] The SEC will inquire if the monetary resources to drill are available and if the producer has a track record of drilling undeveloped locations booked in the past. The locations must meet all regulatory requirements.

If there is not a commitment by the producer to develop the undeveloped reserves, the SEC may disallow the proved classification. The SEC uses a five year time period as a reasonable time frame for development. Lack of commitment can be inferred from the producer's lack of sufficient progress to drill and bring the reserves to market. The producer should have documentation available showing movement toward establishing production, which can include contracting rigs, sales negotiations and building facilities.

Under the December 31, 2008 guidelines, the SEC indicates they do not intend to exclude projects just because they take longer than five years to complete. However, under Item 1203 of Regulation S-K, producers will be required to disclose why the reserves have not been developed.[98]

If outside reserve evaluators are reviewing undeveloped reserves, they will want to see a map showing where the locations are to be drilled. The evaluators will also want assurances from the company they will actually proceed with drilling. The company should have the undeveloped locations on their drilling plans and in their budget.

---

97 SEC Division of Corporation Finance: Frequently Requested Accounting and Financial Reporting Interpretations and Guidance, March 31, 2001 "A commitment by the company to develop the necessary production, treatment and transportation infrastructure is essential to the attribution of proved undeveloped reserves. Significant lack of progress on the development of such reserves may be evidence of a lack of such commitment. Affirmation of this commitment may take the form of signed sales contracts for the products; request for proposals to build facilities; signed acceptance of bid proposals; memos of understanding between the appropriate organizations and governments; firm plans and timetables established; approved authorization for expenditures to build facilities; approved loan documents to finance the required infrastructure; initiation of construction of facilities; approved environmental permits etc. Reasonable certainty of procurement of project financing by the company is a requirement for the attribution of proved reserves. An inordinately long delay in the schedule of development may introduce doubt sufficient to preclude the attribution of proved reserves."

98 Item 1203(d) [17 CFR 229.1203(d)].

Documentation

Undeveloped locations likely will receive more attention from the SEC than producing wells and should be properly documented. All available data should be evaluated before assigning proved reserves to undrilled locations, including analogy to already producing reservoirs in the same geologic section and in the same area. Analogy can be used to estimate reservoir parameters and recovery factors for the undeveloped reservoir. The basis used for the estimate should be available for review by the reserve evaluator or the SEC.

Legal Locations

The SEC will not allow proved reserves if there is not a valid concession or a legal location.[99] Automatic renewal of a concession or granting new spacing rules cannot be assumed unless there is a track record showing that renewals or the granting of new spacing is done as a matter of course.

Reservoir Simulation

The SEC discusses the use of reservoir simulation, material balance and generalized recovery correlations in new reservoirs having only a few wells.[100] They indicate there is too little performance history for a valid history match and the results would be speculative in nature if reservoir simulation is used in a new

---

99 SEC Division of Corporation Finance: Frequently Requested Accounting and Financial Reporting Interpretations and Guidance, March 31, 2001 "The history of issuance and continued recognition of permits, concessions and commerciality agreements by regulatory bodies and governments should be considered when determining whether hydrocarbon accumulations can be classified as proved reserves. Automatic renewal of such agreements cannot be expected if the regulatory body has the authority to end the agreement unless there is a long and clear track record which supports the conclusion that such approvals and renewal are a matter of course."

100 SEC Division of Corporation Finance: Frequently Requested Accounting and Financial Reporting Interpretations and Guidance, March 31, 2001 "In a new reservoir with only a few wells, reservoir simulation or application of generalized hydrocarbon recovery correlations would not be considered a reliable method to show increased proved undeveloped reserves. With only a few wells as data points from which to build a geologic model and little performance history to validate the results with an acceptable history match, the results of a simulation or material balance model would be speculative in nature. The results of such a simulation or material balance model would not be considered to be reasonably certain to occur in the field to the extent that additional proved undeveloped reserves could be recognized. The application of recovery correlations which are not specific to the field under consideration is not reliable enough to be the sole source for proved reserve calculations."

reservoir with only a few wells. The SEC believes recovery correlations, that are not field specific, are not reliable enough to use for proved reserves.

Improved Recovery

Enhanced recovery reserves should not be booked as undeveloped unless there has been a successful test in the area from the same reservoir.

The SEC indicates that proved undeveloped reserves cannot be claimed for improved recovery unless the process has been shown to be effective in the same reservoir or an analogous reservoir.[101] In the December 31, 2008 revision, the SEC expanded the definition to include "other evidence using reliable technology that establishes reasonable certainty".[102] The analogous reservoir must be in the same formation as the improved recovery project but with the new revision, does not have to be in the immediate area[103]. This is an attempt to further align the SEC definitions with the PRMS. An analogous reservoir is defined by the SEC as a reservoir who's geologic and reservoir parameters, such as porosity, permeability, thickness, continuity and hydrocarbon saturation, are as good as or better.

Major Expenditures

The SEC states if a major expenditure is required for a recompletion then the reserves should be classified as undeveloped.[104]

101 SEC Division of Corporation Finance: Frequently Requested Accounting and Financial Reporting Interpretations and Guidance, March 31, 2001 "Reserves cannot be classified as proved undeveloped reserves based on improved recovery techniques until such time that they have been proved effective in that reservoir or an analogous reservoir in the same geologic formation in the immediate area. An analogous reservoir is one having at least the same values or better for porosity, permeability, permeability distribution, thickness, continuity and hydrocarbon saturations."

102 Rule 4-10(a)(25)(iii) [17 CFR 210.4-10(a)(25)(iii)].

103 Securities and Exchange Commission, Modernization of Oil and Gas Reporting, December 31, 2008; p.33.

104 SEC Division of Corporation Finance: Frequently Requested Accounting and Financial Reporting Interpretations and Guidance, March 31, 2001 "(f) Proved undeveloped oil and gas reserves are reserves that are expected to be recovered from new wells on undrilled acreage, or from existing wells where a relatively major expenditure is required for recompletion."

Reliable Technology

The previous SEC rules and guidance allowed for only certain technologies to be used to define proved reserves. The new rules allow the use of "reliable technologies", that is technologies which have been shown to produce reliable and repeatable results in practice.

The SEC defines the term "reliable technology" as field tested technology which has demonstrated it can provide reasonably certain results with consistency and repeatability in the formation being evaluated or an analogous formation[105]. The SEC may have meant analogous reservoirs instead of analogous formations.[106] The SEC has declined to publish a list of acceptable technologies for the determination of proved reserves; instead saying the issuer has the burden of establishing and documenting the technology[107]. This information should be available to the SEC upon request to support the reserve estimates being reviewed.

The SEC recognizes new technologies have developed and will continue to develop making the determination of reserves more reliable. They also say the use of proprietary technologies companies have developed internally will be allowed if they can demonstrate reliable results.[108] If proprietary technology is used, a disclosure of the technology to the SEC staff may be required, but it will not be released to the public.

The required disclosure of reliable technology is limited to a concise summary of the technology used for the estimate. Proprietary technology will not be required

105 17 CFR Parts 210, 211 et al. Modernization of Oil and Gas Reporting; Final Rule, Vol. 74, No 9, II G.1, p. 2166, Jan. 14, 2009.

106 Lee, W. J., Modernization of the SEC Oil and Gas Reserves Reporting Requirements, SPE-123793, p. 8.

107 17 CFR Parts 210, 211 et al. Modernization of Oil and Gas Reporting; Final Rule, Vol. 74, No 9, II G.1, p. 2166, Jan. 14, 2009.

108 Id.

to be disclosed to the public, but rather the disclosure will be more general.[109] An example used by the SEC is combinations of seismic, wireline formation tests, logs and cores used calculate the reserve estimate. This is not to say that the SEC will not request and continue to require a supplemental data to support to the company's conclusion that a technology meets the current definition of reliable technology.

The scientific method requires certain steps[110]. First one must define the question. The next step is to formulate a hypothesis, then perform experiments, interpret the data and draw conclusions. The steps are similar for establishing the application of reliable technology.[111] First define how the technology will help refine the reserve estimate. The next step involves researching the technology and showing how the results have been reliable. One should then show how non-ideal conditions would impact the reliability of the results and the assumptions which must be made. One should include all of the test data in the documentation and keep the analysis updated as new data becomes available.[112]

The SEC does not feel it's necessary to disclose the technology for each filing.[113] The SEC notes the disclosure of the technology used can be general in nature so there is no disclosure of proprietary methodologies. Also, the disclosure will apply to material additions of reserves or reserves reported for the first time based on the technology in question.[114]

The SEC in the December 31, 2008 guidelines note the advances which have been made in technology over the last few years, and especially since the last

---

109 17 CFR Parts 210, 211 et al. Modernization of Oil and Gas Reporting; Final Rule, Vol. 74, No 9, II G.1, p. 2166, Jan. 14, 2009.

110 The Demonstration of a Reliable Technology for Estimating of Oil and Gas Reserves, Aug 27, 2010, Rod Sidle.

111 Id.

112 Id.

113 17 CFR Parts 210, 211 et al. Modernization of Oil and Gas Reporting; Final Rule, Vol. 74, No 9, II G.1, p. 2166, Jan. 14, 2009.

114 Securities and Exchange Commission, Modernization of Oil and Gas Reporting, December 31, 2008; p.37.

definitions were written. They have revised the definition, so any technology, including computational, which has been tested in the field and is consistent and repeatable in the subject formation or an analogous formation can be used. New technologies or combinations of technologies can be used if the company can document its reliability.[115]

Additional Considerations

Consideration should also be given to additional factors where undeveloped locations are being considered, such as the possible depletion of the location by the existing wells and the possibility of having to revise the reserves assigned to the existing wells to account for reserves to be recovered by the locations.[116]

In areas with fractured or otherwise discontinuous, statistical reserves can be assigned if there has been enough wells drilled to establish a trend.[117]

SPE/WPC

The SPE/WPC says undeveloped reserves are undrilled wells, deeper horizons, projects requiring a major expenditure, and facilities or transportation for enhanced recovery projects.

The SPE/WPC allows the use of analogy to assign proved developed reserves for undeveloped enhanced recovery projects.

The SPE/WPC requires reservoir continuity and certainty the project will be

---

115 Securities and Exchange Commission, Modernization of Oil and Gas Reporting, December 31, 2008; p.34.

116 Reserves Definitions Committee Society of Petroleum Evaluation Engineers, "Guidelines for Application of Petroleum Reserves Definitions", p.15.

117 Id.

completed to book proved undeveloped reserves.[118] The proposed location must be a direct offset to a well capable of commercial production and within the known limits of the reservoir. For offsets more than a legal location from a commercial well, the SPE/WPC requires "reasonable certainty" of reservoir continuity as well as commercial accumulations of hydrocarbons at the locations. Reservoir continuity can be based on geologic and engineering data.

The SPE/WPC says offsets more than one legal location from a commercial well are allowed if geological and engineering data indicate with reasonable certainty the reservoir is continuous and has economic reserves.[119] Unlike the SEC which calls for "certainty" of reservoir continuity, the SPE/WPC calls for "reasonable certainty" of reservoir continuity.

Proved oil reserves can be assigned to updip locations if there is a reasonable certainty it is still above the bubble point.

China

The Petroleum Reserve Standard defines proved undeveloped reserves (II or B) as reserves based on "reliable reservoir parameters". They are the basis for the development plan and facilities construction. The error in these reserves should be no more than plus or minus 20%.[120]

---

118 SPE/WPC Reserve Definitions "Reserves in undeveloped locations may be classified as proved undeveloped provided (1) the locations are direct offsets to wells that have indicated commercial production in the objective formation, (2) it is reasonably certain such locations are within the known proved productive limits of the objective formation, (3) the locations conform to existing well spacing regulations where applicable, and (4) it is reasonably certain the locations will be developed. Reserves from other locations are categorized as proved undeveloped only where interpretations of geological and engineering data from wells indicate with reasonable certainty that the objective formation is laterally continuous and contains commercially recoverable petroleum at locations beyond direct offsets."

119 SPE/WPC Reserve Definitions "Reserves from other locations are categorized as proved undeveloped only where interpretations of geological and engineering data from wells indicate with reasonable certainty that the objective formation is laterally continuous and contains commercially recoverable petroleum at locations beyond direct offsets."

120 The National Standard of P.R.C., Petroleum Reserve Standard.

Undeveloped Reserves Narrative Disclosure

With the 2009 SEC Regulations, the SEC requires a company to disclose certain information concerning undeveloped reserves. The items to be disclosed include "...the total quantity of proved reserves at year end"[121]; "...material changes improved undeveloped reserves that occurred during the year, including proved undeveloped reserves converted into proved developed reserves"[122]; "...investments and progress made during the year to convert proved reserves to proved developed reserves, including, but not limited to, capital expenditures"[123]; and "... the reasons why material amounts of proved undeveloped reserves in individual fields or countries remain undeveloped for five years or more after disclosure as proved undeveloped reserves".[124]

"A matter is "material" if there is a substantial likelihood that a reasonable person would consider it important".[125]

The SEC defines undeveloped reserves as "*Undeveloped oil and gas reserves are reserves of any category that are expected to be recovered from new wells on undrilled acreage, or from existing wells where a relatively major expenditure is required for completion.*"[126] They go on to say "*Undrilled locations can be classified as having undeveloped reserves only if a development plan has been adopted indicating that they are scheduled to be drilled within five years, unless the specific circumstances, justify a longer time.*"[127]

121 Federal Register/ Vol 74, No.9 / January 14, 2009 / Rules and Regulations / P. 2195, Part 229.1203.

122 Id.

123 Id.

124 Id.

125 SEC Staff Accounting Bulletin No. 99 – Materiality / Part 211 Section 1.

126 Federal Register/ Vol 74, No.9 / January 14, 2009 / Rules and Regulations / P. 2192, Part 210.4-10 (31).

127 Federal Register/ Vol 74, No.9 / January 14, 2009 / Rules and Regulations / P. 2176, Section IV, 4.

The SEC does not exclude projects which take longer than five years, but require an explanation as to why the reserves have not been developed. They specify several types of projects such as constructing platforms offshore, development in urban areas, remote locations or environmentally sensitive area.[128] Factors to consider in determining whether or not development may be extended past five years include the company's level of significant development activities, their record of completing comparable long-term projects, the amount of time the company has had the lease without significant development activities, the extent to which the company has followed previously adopted development plan, and the extent the delays in development are caused by external factors.[129]

The question has been when the five year time frame starts to run. The answer is it starts when the undeveloped reserves were booked.[130] Previously booked undeveloped locations are not grandfathered. If the project has been booked for five years or more, the reserves should be debooked.[131] If the reserves are not debooked, the circumstances to justify a longer development period should be included.

The question of a development plan for longer than five years has also been raised. The SEC has indicated the five year rule does not apply to probable and possible reserves.[132] No set guideline was mentioned for an appropriate time period. The company must, however, demonstrate intent to develop and have a development schedule in place.[133]

---

128 Question 131.03, SEC Compliance and Disclosure Interpretations Oct. 26, 2009.

129 Id.

130 2010 Ryder Scott Reserves Conference, How Do You Count To Five? "The SEC 5 Year Rule for PUDs Revisited"

131 Id.

132 Id.

133 Id.

A number of factors are considered to determine whether or not a company has committed to a project.[134]

134 RSC SEC Training Module 3, "The Probable and Possible Reserves Dilemma" SEC Website Guidance, "In order to demonstrate commitment to a project, a company must establish a verifiable track record of consistently executing projects as planned. An inordinately long delay in the schedule of development may introduce doubt sufficient to preclude the attribution of proved reserves. Documentation may include, but is not limited to:

- Historical Rig Counts vs. Forecast Rig Schedule
- Historical Capital Spending vs. Capital Budget Forecasts
- Approved authority for expenditures to build facilities
- Appropriate Corporate approvals for funding and execution
- Initiation of construction of facilities
- Signed acceptance of bid proposals
- Requests for proposals to build facilities
- Firm plans and timetables established
- Approved loan documents to finance the required infrastructure
- Reasonable certainty of procurement of project financing
- Signed sales contracts for products
- Memos of understanding between appropriate organizations and governments
- Approved environmental permits"

Volumes in Un-Penetrated Fault Blocks

With the SEC rule changes, there have been many questions, including how to handle potential volumes in un-penetrated fault blocks or reservoirs. For the SEC, the decision involves whether or not the fault is sealing.

One of the SEC requirements for reserve assignment is a known accumulation. The SEC also requires the reservoir to be economically producible.[135] If the undrilled fault block cannot be shown to be connected to the discovered accumulation, the SEC does not consider it a reserve volume.[136] The SEC will consider the volume across a small non-sealing fault as a reserve volume in some circumstances. The SEC will consider if the fault displacement is smaller than the formation thickness and if pressure communication with the known reservoir can be shown. The volume should also be directly adjacent to the known reservoir.[137]

When asked if the un-penetrated pressure separated fault block can be assigned probable or possible reserves, the SEC responded it could not. The answer was an un-penetrated, pressure separated fault block could not be assigned reserves of any category.[138]

The SPE-PRMS defines reserves as commercially recoverable volumes which are discovered. Potentially recoverable volumes and undiscovered volumes are considered prospective resources.

The SPE-PRMS has similar rules as the SEC. The reservoir must be discovered and the volumes commercially recoverable. Much like the SEC, the volumes

135 2010 Ryder Scott Reserves Conference, "What Do You Have If You Don't Have a Well Penetration"

136 Id.

137 Id.

138 Section 117: Rules 4-10(a)(17) and 4-10(a)(18) Definitions -Possible Reserves; Probable Reserves
Question 117.04: Can an issuer assign probable or possible reserves to an unpenetrated fault block?
Answer: No. Un-penetrated, pressure-separated fault blocks should not be considered to contain reserves of any category until penetrated by a well.

must be adjacent to the known volume, across a small non-sealing fault. The fault thrown must be less than the formation thickness and the un-penetrated volume in communication with the known volume.[139]

The SEC and SPE-PRMS treat probable and possible reserves as extensions of established proved areas. They state the volumes are less certain than proved volumes. This may be due to the data quantity and quality or the uncertainty of communication with the known part of the reservoir. To assign reserves to an un-penetrated fault block under either system, one must show reservoir continuity between the known reserves in a penetrated fault block and the un-penetrated fault block.[140] The two fault blocks must be adjacent to each other and the fault must not be sealing. One can use seismic to show the potential fault thickness and reservoir thickness and correlation to seismic amplitudes for hydrocarbon presence.[141]

If the fault is a major sealing fault, the SEC says no reserves of any category can be assigned to the un-penetrated fault block.[142] If one wants to assign reserves across a potentially sealing fault, a pre-booking opinion should be requested from the SEC.[143]

The SPE-PRMS, however, is not as restrictive. The SPE-PRMS indicates caution should be used in assigning reserves to an un-penetrated fault block if the fault is sealing.[144] If one can demonstrate a common hydrocarbon column

139 2010 Ryder Scott Reserves Conference, "What Do You Have If You Don't Have a Well Penetration".

140 Id.

141 Id.

142 Regulation 210.4-10 (a)(26) "Reserves are estimated remaining quantities of oil and gas and related substances anticipated to be economically producible, as of a given date, by application of development projects to known accumulations. ..."Note to paragraph (a)(26):"Reserves should not be assigned to adjacent reservoirs isolated by major, potentially sealing, faults until those reservoirs are penetrated and evaluated as economically producible. ...Such areas may contain prospective resources."

143 2010 Ryder Scott Reserves Conference, "What Do You Have If You Don't Have a Well Penetration".

144 SPE-PRMS Table 3: Reserves Category Definitions, "Caution should be exercised in assigning Reserves to adjacent reservoirs isolated by major, potentially sealing, faults until this reservoir is penetrated and evaluated as commercially productive. Justification for assigning Reserves in such cases should be clearly documented. Reserves should not be assigned to areas that are clearly separated from a known accumulation by non-productive reservoir (i.e., absence of reservoir, structurally low reservoir, or negative test results); such areas may contain Prospective Resources."

and communication, than reserves can be assigned. Some cases include fluid movement up the fault, communication in the geologic past or unconventional reservoirs in which the fault does not segment the accumulation in a broad regional sense.[145]

The SPE-PRMS guidance indicates reserves should not be assigned to areas separated from the known reservoir area by non-productive areas such as no reservoir, low areas or areas with negative test results.[146] These areas preclude the requirement the un-penetrated volume and the known reservoir are adjacent and in communication. A resource classification is more appropriate in these circumstances.[147] The SPE-PRMS classification system has resource categories of Contingent Resources where the risk is commerciality but the volume is discovered and Prospective Resource where the volume is not discovered and commerciality has not been established.[148]

---

145 2010 Ryder Scott Reserves Conference, "What Do You Have If You Don't Have a Well Penetration".

146 SPE-PRMS Table 3: Reserves Category Definitions.

147 Id.

148 2010 Ryder Scott Reserves Conference, "What Do You Have If You Don't Have a Well Penetration".

Basic Proved Reserves

China also has another category of proved reserves called Basic Proved Reserves (III or C).[149] These reserves are used for fractured reservoirs or for fields with multiple producing formations. These reservoirs have a detailed 3-D seismic grid and appraisal wells have been drilled. The reservoir parameters are being acquired with most adequately in place and the reservoir boundaries are "roughly" known. It is the basis for the "rolling exploration and development"[150] phase. During this phase, some development wells are drilled and additional reservoir parameters are obtained. In three years, the reserves can be moved to proved developed after final reserve estimation. The error associated with these reserves is plus or minus 30%.

China allows the water contact to be estimated if there is no contact in the well. It is placed at a point one half the distance between the lowest known hydrocarbon and the top of the next wet sand. The China definitions also allow the use of pressure data to estimate a water contact below the lowest known hydrocarbon.

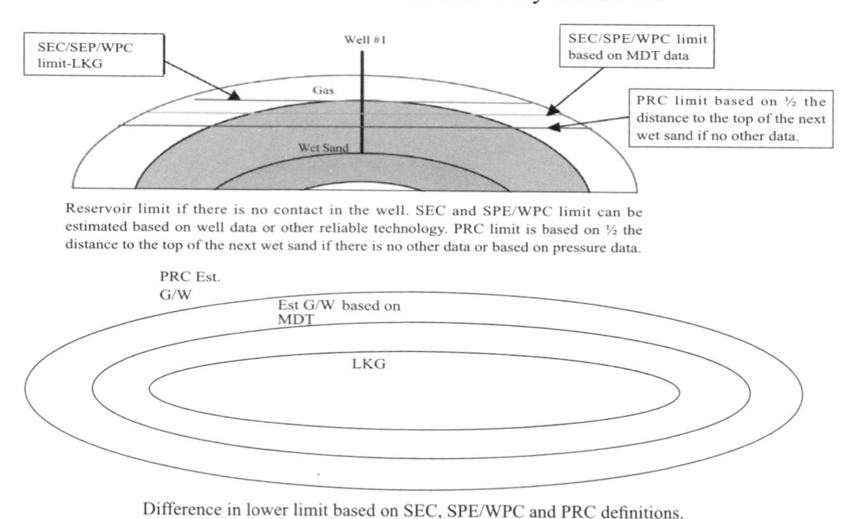

Reservoir limit if there is no contact in the well. SEC and SPE/WPC limit can be estimated based on well data or other reliable technology. PRC limit is based on ½ the distance to the top of the next wet sand if there is no other data or based on pressure data.

Difference in lower limit based on SEC, SPE/WPC and PRC definitions.

149 The National Standard of P.R.C., Petroleum Reserve Standard

150 Id.

Elements of Proved Reserves

Proved reserves under the SEC definitions are:

- quantities of hydrocarbons
- that can be shown with a reasonable certainty
- to be economically recoverable
- under current economic and operating conditions
- from known reservoirs in future years

Quantities of Hydrocarbons

General

The SEC and SPE/WPC, discuss what can be included as "proved reserves". The SEC says "...estimated quantities of crude oil, natural gas, and natural gas liquids."[151] Rule 4-10 applies to "oil and gas activities" which by definition terminate at the first delivery point.[152] They do not include transportation and marketing or

---

151 Rule 4-10(a-2) Regulation S-X, Securities and Exchange Act of 1934, "Proved oil and gas reserves. Proved oil and gas reserves are the estimated quantities of crude oil, natural gas, and natural gas liquids which geological and engineering data demonstrate with reasonable certainty to be recoverable in future years from known reservoirs under existing economic and operating conditions, i.e., prices and costs as of the date the estimate is made. Prices include consideration of changes in existing prices provided only by contractual arrangements, but not on escalations based upon future conditions."

152 Rule 4-10 Regulation S-X, Securities and Exchange Act of 1934, "Oil and gas producing activities.
Such activities include:
The search for crude oil, including condensate and natural gas liquids, or natural gas (oil and gas) in their natural states and original locations.
The acquisition of property rights or properties for the purpose of further exploration and/or for the purpose of removing the oil or gas from existing reservoirs on those properties.
The construction, drilling and production activities necessary to retrieve oil and gas from its natural reservoirs, and the acquisition, construction, installation, and maintenance of field gathering and storage systems—including lifting the oil and gas to the surface and gathering, treating, field processing (as in the case of processing gas to extract liquid hydrocarbons) and field storage. For purposes of this section, the oil and gas production function shall normally be regarded as terminating at the outlet valve on the lease or field storage tank; if unusual physical or operational circumstances exist, it may be appropriate to regard the production functions as terminating at the first point at which oil, gas, or gas liquids are delivered to a main pipeline, a common carrier, a refinery, or a marine terminal.
Oil and gas producing activities do not include:
The transporting, refining and marketing of oil and gas.
Activities relating to the production of natural resources other than oil and gas.
The production of geothermal steam or the extraction of hydrocarbons as a by-product of the production of geothermal steam or associated geothermal resources as defined in the Geothermal Steam Act of 1970.
The extraction of hydrocarbons from shale, tar sands, or coal."

refining.[153] The SPE/WPC definition says "...quantities of petroleum..."[154].

The SPE/WPC do not list specific items which can be reported as reserves but define reserves as quantities of hydrocarbons that are expected to be commercially recovered from a certain date forward.[155]

Martinez[156] says petroleum applies to naturally occurring mixtures of predominately hydrocarbons. Crude oil is hydrocarbon existing in a liquid state in the reservoir and at the surface. Natural gas is the gaseous hydrocarbon phase. It can take the form of solution gas which is dissolved in the liquid phase in the reservoir or associated and non associated gas which is gaseous in the reservoir as well as at the surface. Condensate is a gaseous phase in the reservoir and a liquid at surface conditions. Plant products are liquids extracted from the gas by processing.

Fuel Gas

The SEC has taken the position that produced gas used on the lease as fuel qualifies as proved reserves, but purchased gas would not. The reason given is that the produced gas used as fuel results in a benefit to the company as reduced operating costs. However, fuel gas should not be included under production volumes, or as sales gas, under FASB 69.[157]

---

153 4-10(a-1C) Regulation S-X, Securities and Exchange Act of 1934 "...For purposes of this section, the oil and gas production function shall normally be regarded as terminating at the outlet valve on the lease or field storage tank; if unusual physical or operational circumstances exist, it may be appropriate to regard the production functions as terminating at the first point at which oil, gas, or gas liquids are delivered to a main pipeline, a common carrier, a refinery, or a marine terminal."

154 SPE/WPC Reserve Definitions "Proved reserves are those quantities of petroleum which, by analysis of geological and engineering data, can be estimated with reasonable certainty to be commercially recoverable, from a given date forward, from known reservoirs and under current economic conditions, operating methods, and government regulations. Proved reserves can be categorized as developed or undeveloped."

155 SPE/WPC Reserve Definitions "Reserves are those quantities of petroleum which are anticipated to be commercially recovered from known accumulations from a given date forward."

156 Martinez, A.R. et al. "Classification and Nomenclature Systems for Petroleum and Petroleum Reserves," Proc. 12th World Petroleum Congress, Houston (1987).

157 E-mail from J. Murphy to RECO@SPELIST.SPE.ORG Feb. 8, 2001.

Coalbed Methane

The SEC has indicated that coalbed methane can be included in proved gas reserves. The statement in the prior SEC regulations that gas from coal cannot be included in proved reserves referred to gas processed from coal and not gas occurring in its natural state and original location.[158] Under the new regulations, the SEC allows the disclosure of oil and gas reserves extracted from coal and shale.[159]

Coalbed Methane reserves which are economic only because of federal tax incentives should be noted in the report. The point is that the investors should be made aware of the risks involved.

Gas to Liquid Projects

The SEC will allow reporting of gas, condensate, and natural gas liquids (NLG's), but not other products such as Naphtha and Diesel.

Nontraditional Sources

The SEC allows the disclosure of reserves from nontraditional sources, such as Bitumen and shale as oil and gas reserves. Synthetic oil and gas from Bitumen, shale and coal can also be reported as reserves.[160] The SEC goes on to say a company must include coal and oil shale, if it is to be converted to oil and gas. If it is not to be converted to oil and gas, it should not be included as reserves. The

---

158 Topic 12 of Accounting Series Release No. 257 of the Staff Accounting Bulletins. "Because of a concern over worldwide oil and gas supplies, Congress, in 1980, provided for tax incentives (credits) for the production of oil and gas from other than conventional sources. As a consequence, significant amounts of gas are now recovered from seams of coal beds. This gas is referred to as coalbed methane. It is produced using conventional drilling methods, but for various reasons, it may be more costly to produce than oil and gas recovered from customary sources and some reserves may not be economical without the tax credits.
In instances where methane gas is deemed to be economically producible only as a consequence of existing Federal tax incentives, the staff believes that additional disclosure should be provided as to the specific quantities and values of reported proved reserves that are dependent on existing U.S. tax policy together with any other information necessary to inform readers of the risks attendant with any future change to existing Federal tax policy."

159 **Federal Register** /Vol. 74, No. 9 /Wednesday, January 14, 2009 /Rules and Regulations, p.2163.

160 **Id.**

new SEC regulations have also formalized the guidance for reporting coalbed methane. The SEC has always allowed its inclusion as reserves, but now the issue has been addressed in the regulations. Another possible nontraditional resource is gas hydrates. The SEC states they are shifting the focus of the definition to the final product, regardless of the extraction process.[161]

Volumes Which Are Not Included As Reserves

SEC

The SEC also identifies what proved reserve estimates cannot include.[162] Reserves that are in doubt due to uncertainty in geology, engineering, or economic factors cannot be classified as proved reserves. Another type of reserve that is less than proved is additional reserves that may be recovered from a known reservoir but are classified separately. Hydrocarbons in undrilled fault blocks or reservoirs cannot be shown as proved reserves, but are considered resources under the new SEC regulations.

The definition of "oil and gas producing activities" does not include: "... Transporting, refining, processing (other than field processing of gas to extract liquid hydrocarbons by the company and the upgrading of natural resources extracted by the company other than oil or gas into synthetic oil or gas) or marketing oil and gas;

- The production of natural resources other than oil, gas, or natural resources from which synthetic oil and gas can be extracted;
- The production of geothermal steam"[163]

---

161 **Federal Register** /Vol. 74, No. 9 /Wednesday, January 14, 2009 /Rules and Regulations, p.2163

162 Rule 4-10(a-2iii) Regulation S-X, Securities and Exchange Act of 1934 "Estimates of proved reserves do not include the following: (A) oil that may become available from known reservoirs but is classified separately as "indicated additional reserves"; (B) crude oil, natural gas, and natural gas liquids, the recovery of which is subject to reasonable doubt because of uncertainty as to geology, reservoir characteristics, or economic factors; (C) crude oil, natural gas, and natural gas liquids, that may occur in undrilled prospects; and (D) crude oil, natural gas, and natural gas liquids, that may be recovered from oil shales, coal, gilsonite and other such sources."

163 Securities and Exchange Commission, Modernization of Oil and Gas Reporting, December 31, 2008, p.25.

The SEC does not allow flared gas to be reported as proved reserves.

The SEC has said that income from non oil and gas activities should not be included in reserve reports.[164] These activities include the production and sale of sulphur or CO_2. They also include revenue derived from processing 3rd party gas or other platform operations. These activities can be noted in the Disclosure Section of the 10-K.

Certain items that cannot be reported in the reserve report can be reported in the 10-K Disclosure section. These items could include cash flows with different price scenarios, such as an average price instead of a yearend price. No reserve change should be noted if higher prices are used, but the reserves should be reduced if lower prices are used. Income from non-hydrocarbon sales, such as third party processing fees, can be reported in this section. Material events, such as the acquisition of additional interests in properties in the report or the sale of properties or interests, after the report date cannot be considered in the reserve estimate, but can be reported here. The effect of price hedging can also be disclosed in this section, although it cannot be used in the reserve report.

SPE/WPC

The SPE/WPC does not discuss volumes not included as reserves in the detail of the SEC. They do say, however, gas held in storage should not be reported as reserves.[165] Once gas is removed from its original reservoir, it cannot be counted as reserves if it is re-injected into another reservoir. If it is re-injected into the original reservoir, then it can be counted as reserves until it is produced and sold. The SPE also states reserves can be reduced for usage or losses due to processing.[166]

---

164 SEC engineer cites red flags in reserves reporting; Reservoir Solutions Newsletter, Dec-Feb 2002 p.1.

165 SPE/WPC Reserve Definitions "Reserves do not include quantities of petroleum being held in inventory, and may be reduced for usage or processing losses if required for financial reporting."

166 Id.

Reasonable Certainty

The SPE says the estimation of reserves is done under conditions of uncertainty.[167] The assumptions made in the estimate of reserves should be documented and the reasons given for using those assumptions.[168] The uncertainty involved in reserve estimation is related largely to the amount and reliability of data available and the interpretation of the data. The relative uncertainty is shown by placing the reserves in one of two categories, proved or non-proved.[169]

SEC

All reserve estimates involve some degree of uncertainty.[170] The SEC requires proved reserves be reasonably certain of being recovered, but in the past did not define what reasonable certainty meant. Even if probabilistic methods are used, the SEC did not provide guidance as to what the range of uncertainty should be. The SEC guidance suggests the use of a conservative approach and, with additional data, proved reserve quantities are more likely to go up than down.[171]

---

167 SPE Petroleum Reserves & Resources Definitions

168 Id.

169 SPE Petroleum Reserves & Resources Definitions "All reserve estimates involve some degree of uncertainty. The uncertainty depends chiefly on the amount of reliable geologic and engineering data available at the time of the estimate and the interpretation of these data. The relative degree of uncertainty may be conveyed by placing reserves into one of two principal classifications, either proved or unproved. Unproved reserves are less certain to be recovered than proved reserves and may be further sub-classified as probable and possible reserves to denote progressively increasing uncertainty in their recoverability."

170 SEC Division of Corporation Finance: Frequently Requested Accounting and Financial Reporting Interpretations and Guidance, March 31, 2001. "Probabilistic methods of reserve estimating have become more useful due to improved computing and more important because of its acceptance by professional organizations such as the SPE. The SEC staff feels that it would be premature to issue any confidence criteria at this time. The SPE has specified a 90% confidence level for the determination of proved reserves by probabilistic methods. Yet, many instances of past and current practice in deterministic methodology utilize a median or best estimate for proved reserves. Since the likelihood of a subsequent increase or positive revision to proved reserve estimates should be much greater than the likelihood of a decrease, we see an inconsistency that should be resolved. If probabilistic methods are used, the limiting criteria in the SEC definitions, such as LKH, are still in effect and shall be honored. Probabilistic aggregation of proved reserves can result in larger reserve estimates (due to the decrease in uncertainty of recovery) than simple addition would yield. We require a straight forward reconciliation of this for financial reporting purposes."

171 Securities and Exchange Commission, Modernization of Oil and Gas Reporting, December 31, 2008, p.27.

Under the December 31, 2008 revisions, the term "reasonable certainty" has been defined. The SEC says "reasonable certainty" means the reserves are "much more likely to be achieved than not". The SEC says they believe this definition has the same meaning as the term "high degree of confidence" used in the PRMS and thus have adopted the PRMS standard for deterministic methods.[172] Also in keeping with the PRMS, the SEC definition for probabilistic methods says "there should be at least a 90% probability that the quantities actually recovered will equal or exceed the estimate".[173]

Reasonable certainty requires using valid assumptions based on geological and engineering data. If analogies are used, they should be equal to or better than the reservoir being evaluated. Reasonable certainty should also include financial, operational and political issues.

Uncertainty is related to the amount of reliable data available for the reserve estimate and the ability of the reserve evaluator to interpret that data and make reasonable, supportable assumptions. It also depends on the geologic complexity of the reservoir and the maturity of the project. It depends on the operating conditions and environment for the area and field.

Reasonable certainty is based on the geological and engineering data associated with the reserves assigned, as well as economic, political, legal, and environmental considerations.[174] The assumptions made as to reserves, decline rates, recovery factors and yields must be supported by data and based on a conservative interpretation. If the reservoir is new and data is not available, then the assumptions should be based on a conservative approach. If the drive

172 See Rule 4-10(a)(24) [17 CFR 210.4-10(a)(24)].

173 Securities and Exchange Commission, Modernization of Oil and Gas Reporting, December 31, 2008, p.28.

174 SEC Division of Corporation Finance: Frequently Requested Accounting and Financial Reporting Interpretations and Guidance, March 31, 2001. "The determination of reasonable certainty is generated by supporting geological and engineering data. There must be data available which indicate that assumptions such as decline rates, recovery factors, reservoir limits, recovery mechanisms and volumetric estimates, gas-oil ratios or liquid yield are valid. If the area in question is new to exploration and there is little supporting data for decline rates, recovery factors, reservoir drive mechanisms etc., a conservative approach is appropriate until there is enough supporting data to justify the use of more liberal parameters for the estimation of proved reserves."

mechanism is not known then the most conservative mechanism should be used. In general, the SEC suggests that reserve revisions will more likely be up than down as more data becomes available. [175]

Uncertainty is present in oil and gas reserve estimates. The reserve definitions call for hydrocarbon quantities which are commercially recoverable. There are uncertainties involving the ability of the reservoir to flow the hydrocarbons and how much oil and gas can be recovered from the reservoir. There are uncertainties in determining the pore spaces between the sand grains and the types of fluid filling these voids. The second type of uncertainty involves the prices to be received in the future and the costs to produce the oil and gas. The third type of uncertainty involves political risk in some areas.

An area with little geologic complexity will have less uncertainty than one which is complex. A reservoir on an structure with no faulting and with continuous sands will have a higher degree of certainty than a reservoir that is highly faulted or in a fluvial environment of deposition with discontinuous sands.

A more mature project will have less uncertainty than a relatively new one. As more wells are drilled and put on production, more is known about the reservoir and its producing characteristics. As more production data and pressure history become available, the more certain the reserve estimates become.

The SEC states that uncertainties as to permeability, reservoir continuity, whether or not a fault is sealing, structure, and other reservoir characteristics may prevent the reserves from being classified as proved if they are not known with a reasonable certainty.

In addition, the SEC notes that economic uncertainty can also affect the classification of reserves. There must be a reasonable certainty that a market exists and the reserves will produce a positive cash flow. If there is no definite

175 SEC Division of Corporation Finance: Frequently Requested Accounting and Financial Reporting Interpretations and Guidance, March 31, 2001. "The concept of reasonable certainty implies that, as more technical data becomes available, a positive, or upward, revision is much more likely than a negative, or downward, revision."

market or way to get reserves to market in place or in progress, the reserves cannot be classified as proved. If the reserves are marginal or prices are uneconomic such that the reserves do not have a positive cash flow, they cannot be considered proved.[176]

SPE/WPC

The SPE/WPC[177] says reasonable certainty means a "...high degree of confidence that the quantities will be recovered...". If probabilistic methods are used for the reserve estimate, reasonable certainty is defined as at least 90% probability the reserves will be recovered.

Economically Recoverable

Reserves must generate a positive cash flow when they are produced, meaning revenue exceeds expenses. Any amount of positive cash flow, even as little as $1.00 of undiscounted future net revenue is enough to satisfy the SEC requirements.

If production is shut-in due to prices so low it is uneconomic, the reserves should be removed from the proved category.

The SPE/WPC requires reserves to be commercially producible.[178] If they are not commercially recoverable under existing conditions, they should be placed in the unproved category. The effects of potential changes in economic conditions can be reflected under one of the non-proved categories of reserves such as probable

176 SEC Division of Corporation Finance: Frequently Requested Accounting and Financial Reporting Interpretations and Guidance, March 31, 2001. "Geologic and reservoir characteristic uncertainties such as those relating to permeability, reservoir continuity, sealing nature of faults, structure and other unknown characteristics may prevent reserves from being classified as proved. Economic uncertainties such as the lack of a market (e.g. stranded hydrocarbons), uneconomic prices and marginal reserves that do not show a positive cash flow can also prevent reserves from being classified as proved."

177 SPE/WPC Reserve Definitions, "If deterministic methods are used, the term reasonable certainty is intended to express a high degree of confidence that the quantities will be recovered. If probabilistic methods are used, there should be at least a 90% probability that the quantities actually recovered will equal or exceed the estimate."

178 SPE/WPC Reserve Definitions, "Reserves are those quantities of petroleum which are anticipated to be commercially recovered from known accumulations from a given date forward."

or possible.[179]

China requires the well test at least as good as specified for an industrial rate. Wells producing at lower rates can be produced by local authorities after approval by the Administration for Mineral Reserve of P.R.C. if they can be produced at a profit.

Proved reserves in the PRC, under the new definitions, require the reserves to be economic based on current costs and prices and require production, tests, or analogy to show the formation will produce at economic rates.

Existing Economic and Operating Conditions

General

Existing economic conditions relate to prices, costs, ownership and regulatory requirements in effect on the effective date of the report. The SEC also includes marketing, transportation, and ownership and entitlement terms on the effective date of the report as being a part of the existing economic and operating conditions.[180] The prior SEC guidelines required the use of costs and prices as of the date the report is made. The December 31, 2008 revisions allow for average prices for the year and refer to economic produciblity at current prices and costs. These costs and prices are held constant over the life of the report.

179 SPE/WPC Reserve Definitions "Unproved reserves may be estimated assuming future economic conditions different from those prevailing at the time of the estimate. The effect of possible future improvements in economic conditions and technological developments can be expressed by allocating appropriate quantities of reserves to the probable and possible classifications."

180 SEC Division of Corporation Finance: Frequently Requested Accounting and Financial Reporting Interpretations and Guidance, March 31, 2001. "Existing economic and operating conditions are the product prices, operating costs, production methods, recovery techniques, transportation and marketing arrangements, ownership and/or entitlement terms and regulatory requirements that are extant on the effective date of the estimate. An anticipated change in conditions must have reasonable certainty of occurrence; the corresponding investment and operating expense to make that change must be included in the economic feasibility at the appropriate time. These conditions include estimated net abandonment costs to be incurred and duration of current licenses and permits."

The SPE/WPC[181] requires the use of current economic conditions, but allow the use of historical prices and costs and the use of an averaging period to estimate prices and costs.

Capital Costs and Abandonment Costs

Abandonment costs net of salvage are to be included in the report. Offshore they include plugging the wells, removing the platforms and pipelines and the environmental remediation. As a rule of thumb, abandonment costs are not included in the United States for onshore wells, as the salvage value of the well will generally pay for the well abandonment cost.

If there is a reasonable certainty of an anticipated change in operating conditions, the associated costs should be included, for instance, adding compression or a recompletion.[182]

Capital costs include non-recurring major expenditures that are not covered under lease operating expense. This would include the well cost for undeveloped locations, the recompletion cost for behind pipe reserves, pipeline costs and facilities costs. One could also include major non-recurring expenditures such as major platform maintenance, refurbishing or painting offshore platforms as a capital expense.

---

181 SPE/WPC Reserve Definitions, "Establishment of current economic conditions should include relevant historical petroleum prices and associated costs and may involve an averaging period that is consistent with the purpose of the reserve estimate, appropriate contract obligations, corporate procedures, and government regulations involved in reporting these reserves."

182 SEC Division of Corporation Finance: Frequently Requested Accounting and Financial Reporting Interpretations and Guidance, March 31, 2001. "Existing economic and operating conditions are the product prices, operating costs, production methods, recovery techniques, transportation and marketing arrangements, ownership and/or entitlement terms and regulatory requirements that are extant on the effective date of the estimate. An anticipated change in conditions must have reasonable certainty of occurrence; the corresponding investment and operating expense to make that change must be included in the economic feasibility at the appropriate time. These conditions include estimated net abandonment costs to be incurred and duration of current licenses and permits."

Lease Operating Expense

Operating costs include all of the recurring costs to operate the field. These can include electricity, gathering charges, personnel charges, overhead charges, etc. Typically they are assessed on a per well basis.

In some fields, such as offshore fields or large waterflood projects, a certain percentage of the cost is fixed and not dependent on the number of wells. This cost is allocated to the field as a whole. The variable cost is the remainder of the total operating cost and is allocated to each well. A rule of thumb used in estimating costs on a fixed and variable basis is about 60% to 75% as fixed.

Operating costs should include overhead costs and non-operators must include COPAS in their costs.

Transportation charges are also included as a cost against the net revenue. It can be included as an operating cost, but is generally reported separately.

The SEC has stated operating expenses must be reported as they are incurred.[183] They cannot be reduced by allocating non-hydrocarbon revenue, such as processing fees, against them. The SEC requires the current lease operating expense (LOE) be used and kept constant over the life of the property.

The SPE/WPC says that operating costs can be estimated based on averaging historical costs.[184] As a rule of thumb, monthly costs are averaged for 9 to 12 months to remove any cost spikes and provide a reliable average.

The SEC states that indirect expenses such as overhead and insurance should

183 SEC Division of Corporation Finance: Frequently Requested Accounting and Financial Reporting Interpretations and Guidance, March 31, 2001. "Existing economic and operating conditions are the product prices, operating costs, production methods, recovery techniques, transportation and marketing arrangements, ownership and/or entitlement terms and regulatory requirements that are extant on the effective date of the estimate. An anticipated change in conditions must have reasonable certainty of occurrence; the corresponding investment and operating expense to make that change must be included in the economic feasibility at the appropriate time."

184 SPE/WPC Reserve Definitions "Establishment of current economic conditions should include relevant historical petroleum prices and associated costs and may involve an averaging period that is consistent with the purpose of the reserve estimate, appropriate contract obligations, corporate procedures, and government regulations involved in reporting these reserves."

be allocated to the proved reserves using reasonable methods.[185] For the non-operator, the COPAS charges must be included, but the operator is not allowed to count these as revenue or reduce their operating expense based on this revenue.

The SEC will audit the operating expense reported by comparing the first year expense of the proved producing reserves to the operating expense in the prior years 10-K.[186]

The PRC, under the new definitions, requires using the costs on the report date or as specified by contract for proved reserves.

Taxes

Taxes, like operating costs, must be subtracted from the revenue received from the sale of hydrocarbons. The two most common in the United States are the severance tax and the as-valorem tax.

Severance taxes are levied on the produced oil and gas by the individual states in the United States. The vary in amount but are generally related to a percentage of revenue. They may also vary with production rates.

Ad-valorem taxes are levied by the counties in each state. They can be based on the value of the production or the equipment. They may be paid by the operator and included in the operating costs.

Facilities

The SEC requires that facilities, pipelines, compressors, gathering lines be installed or construction contracts and permits be in place. If the facilities are to be installed later, such as compression, the cost for such should be included and it should be indicated it is a delayed installation.

---

185 SEC engineer cites red flags in reserves reporting; Reservoir Solutions Newsletter, Dec-Feb 2002, p.1.

186 Id.

Prices

Prior to the December 31, 2008 SEC guidelines, the prices used for reporting future revenue for SEC filings were year-end prices and the SEC required the actual price received on the last day of the year to be used for year end reports or the effective date of the report for other reporting dates. This requirement was the result of FASB 69 which was promulgated in 1982. The producer was to take the posted price for the last day of the year and apply historic adjustments such as transportation, gravity adjustments, etc. and use this on an individual property basis in his report.[187] Although the monthly average price may be what the company actually receives, the SEC would accept only the actual price on the last day of the year.

With the December 31, 2008 Guidelines, the SEC revised the price requirements, such that the reports are to use 12 month average prices. The prices are calculated using the unweighted arithmetic average of the first day of the month price for each of the 12 months prior to the end of the reporting period.[188] The SEC notes the 12 month average price is thought by some to be comparable to the average price used by the PRMS.[189] The SEC feels the first day of the month prices will allow the reporting companies more time to prepare the price estimate.[190]

Spot oil prices are not posted on weekends or holidays, and posted oil prices are in effect until changed. If the first day of the month is on a weekend or holiday, the last preceding price is used for the first day of the month price. For example, if the first day of the month is on a Sunday, then the preceding posted Friday

187 Statement of Financial Accounting Standards 69, paragraph 30.a. "Future cash inflows . . . be computed by applying year-end prices of oil and gas relating to the enterprise's proved reserves to the year-end quantities of those reserves. This requires the use of physical pricing determined by the market on the last day of the (fiscal) year. For instance, a west Texas oil producer should determine the posted price of crude (hub spot price for gas) on the last day of the year, apply historical adjustments (transportation, gravity, BS&W, purchaser bonuses, etc.) and use this oil or gas price on an individual property basis for proved reserve estimation and future cash flow calculation (this price is also used in the application of the full cost ceiling test). A monthly average is not the price on the last day of the year, even though that may be the price received for production on the last day of the year."

188 See Rule 4-10(a)(22)(v) [17 CFR 210.4-10(a)(22)(v)].

189 Securities and Exchange Commission, Modernization of Oil and Gas Reporting, December 31, 2008, p.12.

190 Securities and Exchange Commission, Modernization of Oil and Gas Reporting, December 31, 2008, p.21.

price is used as the price for the first day of the month.

Gas is priced differently than oil. It is priced the day before the sale date. It is also different since it is priced on weekends and holidays. Gas sold on the first day of the month is based on a price set the previous day.

The SEC requires the assumption of continuing current economic conditions.[191] That is, the future revenue is to be estimated using current prices and costs with no escalations. The SEC does make an exception if the product is subject to a sales contract with definite, definable increases that are not tied to inflation.

As with the old definitions, the new December 31, 2008 definitions specify that if price is subject to a contract, then the price specified in the contract, along with any changes specified, is used in the estimate of future net revenues.[192] After the contract expires, the price to be used is the current market price at the end of the fiscal year. [193]

The SEC requires that fixed prices be used for all purposes including economic limits, future revenue, and PSC/PSA reserves.

If gas is not subject to a sales contract, the SEC requires the current market price of similar gas be used.[194] Price increases announced but not yet effective should

---

191 Statement of Financial Accounting Standards 69, paragraph 30.b, "... future production costs are to be based on year-end figures with the assumption of the continuation of existing economic conditions."

192 Rule 4-10(a)(22)(v) [17 CFR 210.4-10(a)(22)(v)].

193 Topic 12 of Accounting Series Release No. 257 of the Staff Accounting Bulletins "Question 1: For purposes of determining reserves and estimated future net revenues, what price should be used for gas which will be produced after an existing contract expires or after the redetermination date in a contract? Interpretive Response: The price to be used for gas which will be produced after a contract expires or has a redetermination is the current market price at the end of the fiscal year for that category of gas. This price may be increased thereafter only for additional fixed and determinable escalations, as appropriate, for that category of gas. A fixed and determinable escalation is one which is specified in amount and is not based on future events such as rates of inflation."

194 Topic 12 of Accounting Series Release No. 257 of the Staff Accounting Bulletins "Question 2: What price should be applied to gas which at the end of a fiscal year is not yet subject to a gas sales contract? Interpretive Response: The price to be used is the current market price for similarly situated gas at the end of the fiscal year provided the company can reasonably expect to sell the gas at the prevailing market price."

not be used as year end prices for SEC reports.[195]

The SEC requirement to use the actual price on the last day of the year has caused significant reserve revisions in the past. At the time the regulation was written, more gas was sold on long term contracts, while today, spot market sales are more common. The SEC feels that the average monthly price in the December 31, 2008 revisions are more consistent with the objective of being able to compare various companies.[196]

The SEC will allow hedge transactions only to the extent they are property specific, which would be a rare occurrence. If hedge transactions result in significant positive or negative differences in the future net revenue and reserves, they should be noted in the disclosure section of the report.

The PRC, under the new definitions, requires the reserves to be economic under costs and prices on the report date or based on contract. Under the new SEC guidelines, the economic producibility must be based on current economic conditions. The SEC indicates the disclosure rules are not intended to represent the fair market value of reserves, but to rather the relative amount likely to be recovered using a method which limits the impact of non reserve factors.[197]

Alternate Pricing

The Disclosure section can also be used to report alternate cash flows if they are based on reasonable assumptions as to future prices. Reserves resulting from these alternative cash flows should not increase if higher prices are assumed, but should be lowered if lower prices are assumed.

---

195 Topic 12 of Accounting Series Release No. 257 of the Staff Accounting Bulletins "Question 3: To what extent should price increases announced by OPEC or by certain government agencies not yet effective at the date of the reserve report be considered in determining current prices? Interpretive Response: Current prices should not reflect price increases announced but not yet effective at the date of the reserve valuation, i.e., the end of the fiscal year."

196 Securities and Exchange Commission, Modernization of Oil and Gas Reporting, December 31, 2008, p.14.

197 Securities and Exchange Commission, Modernization of Oil and Gas Reporting, December 31, 2008, p.19.

Miscellaneous Issues

Flow Tests

The SEC requires confirmation the reservoir can produce economically.[198] The SEC uses the term "conclusive formation test" and has said that this is more than a small wireline sample or a test of a "few hundred barrels per day" in a remote area. The SEC, with the December 31, 2008 revision, allows the use of any reliable technology which has been tested in the field and demonstrated repeatability in the formation of interest or an analogous formation.[199]

The SEC will accept proved reserves without a flow test if a combination of log and core data indicate productivity of a reservoir and it is analogous to a similar reservoir if the field that has produced or been flow tested.

The SPE/WPC requirement for economic production is similar to the SEC's. The likelihood of commercial production should be shown by actual production or a formation test.[200]

The SPE/WPC allows proved reserves to be based on logs and cores if they are analogous to wells tested or producing in the same area. [201]

China has established minimum rates for a well to be considered proved or to have industrial flow.[202] If the well tests below these rates, reserves are not estimated. For reservoirs at less than 500 m, it must test 0.3 t/d of oil or 500 cubic meters of gas per day. For reservoirs of 500 to 1000 m the rates increase to

---

198 Rule 4-10(a-2i) Regulation S-X, Securities and Exchange Act of 1934 "Reservoirs are considered proved if economic producibility is supported by either actual production or conclusive formation test."

199 Securities and Exchange Commission, Modernization of Oil and Gas Reporting, December 31, 2008, p.34.

200 SPE/WPC Reserve Definitions "In general, reserves are considered proved if the commercial producibility of the reservoir is supported by actual production or formation tests. In this context, the term proved refers to the actual quantities of petroleum reserves and not just the productivity of the well or reservoir. In certain cases, proved reserves may be assigned on the basis of well logs and/or core analysis that indicate the subject reservoir is hydrocarbon bearing and is analogous to reservoirs in the same area that are producing or have demonstrated the ability to produce on formation tests."

201 Id.

202 The National Standard of P.R.C., Petroleum Reserve Standard

0.5 t/d of oil or 1000 cubic meters of gas per day. For 1000 to 2000 m the rates are 1t/d of oil or 3000 cubic meters of gas. For 2000 to 3000 m in depth, the rates are 3 t/d of oil or 5000 cubic meters of gas. At 3000 to 4000 m it is 5 t/d of oil or 10,000 cubic meters of gas. Any reservoir deeper than 4000 m must test at 10 t/d of oil or 20,000 cubic meters of gas.

Use of Data Obtained After Year End

The SEC requires reserves to be based on data received up to the report date and not data received afterwards. If data received after the report date materially affects the stated reserves, the evaluator should mention the change in the text of the report. Events that would affect the reserves include purchase or sale of producing properties, dry holes where undeveloped locations have been set up, failure of facilities, etc.

Ownership

Hydrocarbon interests are generally described as working interests and net revenue interests. The net revenue interest is the working interest minus any royalties. The working interest is the interest used to calculate the net costs while the net revenue interest is used to calculate the net revenue from the sale of hydrocarbons.

In many countries, including the PRC, ownership of the minerals is with the government. Producers receive the right to extract the hydrocarbons, but have no ownership in them. These interests are governed by a production sharing agreement between the host country and the producer. The SEC requires that these volumes are reported separately from mineral interest volumes.

According to the SEC, an oil and gas company should report hydrocarbon volumes net to their leasehold interest.[203]

203 Topic 12 of Accounting Series Release No. 257 of the Staff Accounting Bulletins "Companies should report reserves of natural gas liquids which are net to their leasehold interests..."

PSC/PSA Volumes

With a PSC, the government usually retains ownership of the hydrocarbons. The production company assumes the risk of exploration and drilling, and after a discovery, a contract is written whereby the company recovers their cost out of production and receives a share of the profits, which may be taken in kind. The company's reserve share of the profits can be determined based on their working interest net of royalty, or by taking cost recovery revenue and the profit revenue and dividing it by the year end price for their barrel entitlement. This latter method is the economic recovery method and is preferred by the SEC.[204]

The SEC has expressed concerns about reporting reserves associated with PSC's. The SEC wants to separate reserves owned as leasehold interests and reserves owned through PSC's controlled by foreign governments. Reserves related to PSC's should be listed separately in the report and it should be stated which reserves are for PSC's and which are for mineral interests.

The SEC does not require that the production company has the right to take production in kind, but the company must have the right to extract hydrocarbons and they must have capital at risk. The SEC recognizes a company's legal right to produce after the foreign government has issued a Declaration of Commerciality or has approved the Development Plan. Exceptions are made only if a compelling case is made to the SEC. The company should not project and book reserves beyond the life of the PSC.

---

204 SEC Division of Corporation Finance: Frequently Requested Accounting and Financial Reporting Interpretations and Guidance, March 31, 2001 3(I) "... In general, two methods of determining oil and gas reserves under production sharing arrangements have been proposed by registrants: (a) the working interest method and (b) the economic interest method. Under the working interest method, the estimate for total proved reserves is multiplied by the respective working interest held by the contracting company, net of any royalty. Under the economic interest method, the company's share of the cost recovery oil revenue and the profit oil revenue is divided by the year-end oil price, which represents the volume entitlement. The lower the oil price, the higher the barrel entitlement, and vice versa.
Reserve volumes determined by various owners should add up to 100% of the total field reserves, but that is not always the case using the working interest method. If the working interest is different from the profit entitlement, the economic interest method is the method acceptable to the staff because it is a closer representation of the actual reserve volume entitlement that can be monetized by a company. Also, use of the economic interest method avoids violating the prohibition in paragraph 10 of SFAS 69 against reporting reserves owned by others."

Net Profits Interests

The handling of a net profits interest (NPI) has also been addressed by the SEC. The SEC guidelines indicate that the NPI should not be handled as an increase in lease operating expense. The production and revenues reported must be net of any NPI, and reserves must be reduced by the NPI amount.

The SEC requires that the NPI reserves be deducted from the reserves owned by the interest subject to the NPI.

Plant Volumes

In the case of plant volumes, the SEC indicates the company should report the portion of the plant recovery allocated to the leasehold interest.[205]

Processing Fees

The SEC, in the October 22, 2002 SPEE Forum, said third party processing fees cannot be included as income since they are not oil and gas revenues as defined by Rule 4-10. It is possible these revenues can be included in a different section of the SEC filing, but should not be in the reserve section. The processing expense, however, should be included as a cost against the reserve revenue. The SEC has applied this rational even if the same party owns the producing platform and the receiving platform.

If the additional expense for third party gas production related to operating the processing facilities can be broken out and identified separately, than this additional expense, beyond what it would cost to only produce the company's reserves, can possibly be deducted from the operating expense. The company

205 Topic 12 of Accounting Series Release No. 257 of the Staff Accounting Bulletins "Question 3: What volumes of natural gas liquids should be reported as net reserves, that portion recovered in a gas processing plant and allocated to the leasehold interest or the total recovered by a plant from net interest gas? Interpretive Response: Companies should report reserves of natural gas liquids which are net to their leasehold interests, i.e., that portion recovered in a processing plant and allocated to the leasehold interest. It may be appropriate in the case of natural gas liquids not clearly attributable to leasehold interests ownership to follow instructions to Item 3 of Securities Act Industry Guide 2 and report such reserves separately and describe the nature of the ownership."

needs to be able to show what portion of the operating expense is due to processing the additional third party hydrocarbons. The deduction must be defendable and not an arbitrary reduction.

Foreign Concessions

Foreign concessions are generally governed by production sharing contracts (PSCs). The duration of the license should be considered and reserves not assigned beyond the life of the PSC.

Before proved reserves are assigned in a foreign concession, there should be an approved plan of development (POD) and established markets.

The SEC requires government approval of the POD and a market before they consider reserves proved.

Market Availability

The SEC requires availability of a market for the hydrocarbons. Although a signed contract may not be required, there should at least be a memorandum of understanding which spells out the terms of the agreement and the price to be received for.[206]

Recovery Factors

The SEC may require companies to provide support for the recovery factors used in their estimate of recoverable reserves. They want to see recovery factors on the low side, i.e. the more unfavorable anticipated drive mechanism, until there is enough production data to decide the drive mechanism. Analogy to other production from the same formation can be used if the current reservoir is as good as or better than the analogy. The assumptions used for residual gas

206 SEC Division of Corporation Finance: Frequently Requested Accounting and Financial Reporting Interpretations and Guidance, March 31, 2001 3(d) "...In developing frontier areas, the existence of wells with a formation test or limited production may not be enough to classify those estimated hydrocarbon volumes as proved reserves. Issuers must demonstrate that there is reasonable certainty that a market exists for the hydrocarbons and that an economic method of extracting, treating and transporting them to market exists or is feasible and is likely to exist in the near future."

saturation, residual oil saturation, sweep efficiency, and abandonment pressure must be reasonable and within the expected range of values. There must be documentation to support the assumptions relied on for the recovery factor estimate.

In China, oil reservoirs are usually assigned water drive recovery factors since they typically require pressure support either through water injection or gas injection to produce effectively.

Future production Rates

Future production rates can be based on decline curves, tests, models or analogy.[207] The reserve report should state the method used to estimate the future production rates and any unusual situations.

Consideration should be given to the type of reservoir and the drive mechanism anticipated. A tight depletion drive reservoir will produce differently than a high permeability water drive reservoir.

Initial rates, in the absence of production, should reflect the test rates, any regulatory limitations, capacity of the production facilities, and sales contracts.[208] The structural location of the well should also be considered. If the well is near the contact, a lower rate should be considered.

If there is production history, the rates should reflect the actual rates being produced.[209] The rates should also consider fluctuations in the market.

Disproportionate Value

The SEC will look closely at a field if the company has only one field as

207 Reserves Definitions Committee Society of Petroleum Evaluation Engineers, "Guidelines for Application of Petroleum Reserves Definitions", p.33.

208 Id.

209 Reserves Definitions Committee Society of Petroleum Evaluation Engineers, "Guidelines for Application of Petroleum Reserves Definitions", p.34.

the major part of its reserve base. The SEC will also pay more attention to companies if a disproportionate part of their reserve base is non-producing.

Reservoir Area

The reservoir area is restricted by the SEC to the area defined by drilling and limited downdip by water contacts. In the absence of an observed water contact, the SEC bases the lower limit on the lowest known hydrocarbons as defined by logs or flow tests in the well.[210] The SEC also includes areas adjacent to the drilled portion of the reservoir if there is a reasonable presumption of economic reserves based on the available geological and engineering data.

The SEC has been unyielding on the lowest known hydrocarbon as the reservoir limit when there is no water contact. The lowest known occurrence must be based on hard data from the well bore such as wireline logs. To date they have not allowed the extension of the reservoir area below the lowest known hydrocarbon seen in the wellbore based on MDT data or 3-D seismic data and amplitudes.

Upon obtaining performance history sufficient to reasonably conclude that more reserves will be recovered than those estimated volumetrically down to LKH, positive reserve revisions should be made.[211]

The SEC has indicated a pressure limits test cannot be used to determine the gas/ water contact. They state this will make a good case for probable reserves but it is not adequate for proved reserves as there is still no data for the reservoir

---

210 Rule 4-10(a-2i) Regulation S-X, Securities and Exchange Act of 1934 "The area of a reservoir considered proved includes (A) that portion delineated by drilling and defined by gas-oil and/or oil-water contacts, if any; and (B) the immediately adjoining portions not yet drilled, but which can be reasonably judged as economically productive on the basis of available geological and engineering data. In the absence of information on fluid contacts, the lowest known structural occurrence of hydrocarbons controls the lower proved limit of the reservoir."

211 SEC Division of Corporation Finance: Frequently Requested Accounting and Financial Reporting Interpretations and Guidance, March 31, 2001 "Upon obtaining performance history sufficient to reasonably conclude that more reserves will be recovered than those estimated volumetrically down to LKH, positive reserve revisions should be made."

parameters below the lowest known hydrocarbon.[212]

If a volumetric estimate has been limited to the lowest known hydrocarbon, it can be increased as more performance data becomes available.[213]

If an acreage assignment is made due to the lack of data to adequately define the reservoir, 40 acres for oil and 160 acres for gas seems acceptable to the SEC in most cases. This may not be true, however, in areas such as salt domes or reefs where the reservoir area tends to be small.

The SPE/WPC guidance on reservoir area is similar to the SEC's.[214] It includes the area defined by drilling and fluid contacts. Like the SEC the lower limit of the reservoir is based on fluid contacts, or if no contact, then the lowest occurrence of hydrocarbons. The SPE/WPC bases the lowest occurrence on logs or tests, but not seismic.

The SPE/WPC unlike the SEC allows the lower limit of the reservoir to be estimated by means other than the wireline log. The SPE/WPC comment saying the lower limit can be based on engineering data would imply MDT data can be used to estimate a contact below the lowest known hydrocarbon on the wireline log.

The PRC delineates the reservoir area based on a structure map on the top of the

212 Data from pressure limits tests are not compelling enough to prove up reserves below LKH, says SEC, Reservoir Solutions Newsletter June-August 2003 "In response to the proposed use of pressure-limits tests to push the proved contact below the lowest known gas from the log data, one SEC engineer stated that he would be "...very uncomfortable with doing that based on seismic and pressure-transient analysis. "He reasoned that although an evaluator may be able to determine the contact, he still has no data on the reservoir below the lowest known gas, including saturations, porosity and net pay.
The SEC engineer said that the presenter demonstrated a very good method of determining probable reserves but that this case does not rise to the level of certainty of proved reserves by SEC standards."

213 SEC Division of Corporation Finance: Frequently Requested Accounting and Financial Reporting Interpretations and Guidance, March 31, 2001 "Upon obtaining performance history sufficient to reasonably conclude that more reserves will be recovered than those estimated volumetrically down to LKH, positive reserve revisions should be made."

214 SPE/WPC Reserve Definitions "The area of the reservoir considered as proved includes (1) the area delineated by drilling and defined by fluid contacts, if any, and (2) the undrilled portions of the reservoir that can reasonably be judged as commercially productive on the basis of available geological and engineering data. In the absence of data on fluid contacts, the lowest known occurrence of hydrocarbons controls the proved limit unless otherwise indicated by definitive geological, engineering or performance data."

reservoir using logs, seismic and tests. Proved reservoir limits can be delineated based on boundary tests or by pressure tests if there is not a water contact. If the boundary is a fault, it should be penetrated by more than one well.

China allows booking proved reserves below the low known hydrocarbons. The China standard is one half the distances from the low known hydrocarbon to the next perforated non-productive sand.

Effective Thickness

The SEC limits the net sand thickness to the net sand seen in the wellbore and do not allowed thickening of the sand based on seismic control. Effective thickness is defined as that thickness that has moveable hydrocarbons in the National Standard of PRC.[215] It is the hydrocarbon bearing thickness less the non-hydrocarbon rock. Cores and tests should be used to define the effective thickness. Petrophysical data, such as porosity, permeability and movable oil, should be based on cores and correlated to core data. The minimum thickness for reservoir beds should be in the range of 0.2 to 0.4 meter and of the shale interbeds, 0.2 meter.[216] When the zone has been tested at the minimum rate specified by the standard industrial flow, it can be included as effective thickness. For proved reserves, the logs should be compared to the cores when available. The cutoffs for porosity, water saturation and permeability can be based on cores, tests or analogy.

Improved Recovery

Additional reserves based on improved recovery can be classified as proved according to the SEC definitions, if there is a successful pilot or the project is

215 The National Standard of P.R.C., Petroleum Reserve Standard.

216 Id.

ongoing and performance indicates the additional reserves are justified.[217] The company may use an ongoing project in the same formation in the immediate area as an analogy to assign proved reserves if the formation properties in the new project are as good as or better than in the analogy.[218]

The SEC makes a distinction between pressure maintenance projects where injection is started early in the life of the project and secondary recovery projects.[219] They have stated that in limited situations, such as some North Sea fields, the distinction can be made. The SEC will review the data presented by the company to support the contention improved recovery can actually be achieved.

The SPE/WPC definitions require a successful pilot or analogy to a successful project in an analogous reservoir with similar rock and fluid properties for the assignment of reserves based on improved recovery.[220] The SPE/WPC definitions

217 Rule 4-10(a-2ii) Regulation S-X, Securities and Exchange Act of 1934 "Reserves which can be produced economically through application of improved recovery techniques (such as fluid injection) are included in the "proved" classification when successful testing by a pilot project, or the operation of an installed program in the reservoir, provides support for the engineering analysis on which the project or program was based."

218 SEC Division of Corporation Finance: Frequently Requested Accounting and Financial Reporting Interpretations and Guidance, March 31, 2001 3(c) "...If an improved recovery technique which has not been verified by routine commercial use in the area is to be applied, the hydrocarbon volumes estimated to be recoverable cannot be classified as proved reserves unless the technique has been demonstrated to be technically and economically successful by a pilot project or installed program in that specific rock volume. Such demonstration should validate the feasibility study leading to the project."

219 Topic 12 of Accounting Series Release No. 257 of the Staff Accounting Bulletins Question 2: In determining whether "proved undeveloped reserves" encompass acreage on which fluid injection (or other improved recovery technique) is contemplated, is it appropriate to distinguish between (i) fluid injection used for pressure maintenance during the early life of a field and (ii) fluid injection used to effect secondary recovery when a field is in the late stages of depletion? The definition in Rule 4-10(a)(4) does not make this distinction between pressure maintenance activity and fluid injection undertaken for purposes of secondary recovery.
Interpretive Response: The Office of Engineering believes that the distinction identified in the above question may be appropriate in a few limited circumstances, such as in the case of certain fields in the North Sea. The staff will review estimates of proved reserves attributable to fluid injection in the light of the strength of the evidence presented by the registrant in support of a contention that enhanced recovery will be achieved.

220 SPE/WPC Reserve Definitions "Reserves which are to be produced through the application of established improved recovery methods are included in the proved classification when (1) successful testing by a pilot project or favorable response of an installed program in the same or an analogous reservoir with similar rock and fluid properties provides support for the analysis on which the project was based, and, (2) it is reasonably certain that the project will proceed. Reserves to be recovered by improved recovery methods that have yet to be established through commercially successful applications are included in the proved classification only (1) after a favorable production response from the subject reservoir from either (a) a representative pilot or (b) an installed program where the response provides support for the analysis on which the project is based and (2) it is reasonably certain the project will proceed."

also include the statement there must be a reasonable certainty the project will proceed.

Frontier Areas

In frontier areas, there are additional factors the SEC considers in addition to having a well with a commercial test.[221] There must be a market and a way to get reserves to market or a plan in place that a market will exist in the near future. The company must show that they can economically produce, process, transport and market the hydrocarbons.

There must be a commitment by the company to develop the production, treatment and transportation infrastructure to produce and sell the reserves.[222] A lack of progress indicates to the SEC a lack of commitment by the company and the reserves should not be classified as proved. Evidence of progress can include signed sales contracts, proposal requests for facilities, memorandum of understanding with the appropriate governments, firm plans and timetables, an approved authority for expenditure for facilities, approved loan documents,

---

221 SEC Division of Corporation Finance: Frequently Requested Accounting and Financial Reporting Interpretations and Guidance, March 31, 2001 3(d) "...In developing frontier areas, the existence of wells with a formation test or limited production may not be enough to classify those estimated hydrocarbon volumes as proved reserves. Issuers must demonstrate that there is reasonable certainty that a market exists for the hydrocarbons and that an economic method of extracting, treating and transporting them to market exists or is feasible and is likely to exist in the near future. A commitment by the company to develop the necessary production, treatment and transportation infrastructure is essential to the attribution of proved undeveloped reserves. Significant lack of progress on the development of such reserves may be evidence of a lack of such commitment. Affirmation of this commitment may take the form of signed sales contracts for the products; request for proposals to build facilities; signed acceptance of bid proposals; memos of understanding between the appropriate organizations and governments; firm plans and timetables established; approved authorization for expenditures to build facilities; approved loan documents to finance the required infrastructure; initiation of construction of facilities; approved environmental permits etc. Reasonable certainty of procurement of project financing by the company is a requirement for the attribution of proved reserves. An inordinately long delay in the schedule of development may introduce doubt sufficient to preclude the attribution of proved reserves..."

222 SEC Division of Corporation Finance: Frequently Requested Accounting and Financial Reporting Interpretations and Guidance, March 31, 2001. 3(d) "... A commitment by the company to develop the necessary production, treatment and transportation infrastructure is essential to the attribution of proved undeveloped reserves. Significant lack of progress on the development of such reserves may be evidence of a lack of such commitment. Affirmation of this commitment may take the form of signed sales contracts for the products; request for proposals to build facilities; signed acceptance of bid proposals; memos of understanding between the appropriate organizations and governments; firm plans and timetables established; approved authorization for expenditures to build facilities; approved loan documents to finance the required infrastructure; initiation of construction of facilities; approved environmental permits etc. Reasonable certainty of procurement of project financing by the company is a requirement for the attribution of proved reserves."

initiation of facility construction, and environmental permits. A long delay in development may be enough to keep the reserves from being classified as proved. The SEC does not state a length of time it considers as unreasonable, but looks at each case individually.

The SEC requires a reasonable of certainty project financing before reserves are classified as proved.[223]

223 SEC Division of Corporation Finance: Frequently Requested Accounting and Financial Reporting Interpretations and Guidance, March 31, 2001 "Reasonable certainty of procurement of project financing by the company is a requirement for the attribution of proved reserves."

Volumetric Method of Reserve Estimation

General

The volumetric method of estimating reserves is a static reserve estimate. It is of particular use early in the life of the field development when there is little performance data available. It should also be reviewed throughout the life of the field as a check against the dynamic methods used.

The volumetric method is based on the equation

$$\text{Volume in place} = \frac{(A*h) * phi * (1-Sw)}{Bo}$$

where

A = Area

h = Net thickness

phi = porosity

Sw = water saturation

Bo (Bg for gas) = Formation volume factor

Reserves are then equal to volume in place * recovery factor.

The SEC and the SPE/WPC do not elaborate on how reservoir parameters should be determined. The procedures and limits are not spelled out as such in either set of definitions.[224] There are however certain accepted industry standards used to characterize these parameters. China and Canada do, however, delineate the methods to be used. This paper will touch on the industry standards and the procedures outlined by China.

224 Auditing Standards for Reserves, SPE, June 2001 5.4 "Estimating Reserves by the Volumetric Method Estimating reserves in accordance with the volumetric method involves estimation of oil in place based upon review and analysis of such documents and information as (i) ownership and development maps; (ii) geologic maps; (iii) electric logs and formation tests; (iv) relevant reservoir and core data; and (v) information regarding the completion of oil and gas wells and any production performance thereof. An appropriate estimated recovery efficiency is applied to the resulting oil in place figure in order to derive estimated reserves."

The SPE does say that there is uncertainty in any type of estimate. The SPE defines uncertainty as a range of possible outcomes for an estimate.[225] In general, the more data available, the less uncertain the estimate is. The SPE also describes the main uncertainties associated with a volumetric estimate. These include uncertainty about the reservoir geometry, pore volume, permeability distribution, fluid contacts, and various combinations of reservoir quality.[226]

An oil and gas reservoir is a porous stratum which can contain oil, gas, or water or a combination. For the fluids to flow, the strata must also have permeability.[227]

The steps in a volumetric estimate include estimating the net pay, rock and fluid properties, reservoir limits and structural and isopach maps. To determine net pay, one must establish cutoff values for porosity, water saturation and shale volume. The rock and fluid properties need to be described as accurately as possible. The maps need to fit a geologic model appropriate for the reservoir in question. The maps need to be mechanically correct. The maps need to have adequate legends and scales. A bar scale is preferred.

The hydrocarbon in place volume depends on the amount of pore space or the volume available to hold the reservoir fluid, whether the fluid is water or hydrocarbon. Porosity is the measure of total volume while hydrocarbon saturation is the relative amount of the porosity that is hydrocarbon bearing. The remainder of the pore space contains water.

---

225 Petroleum Resources Management System 2007 Sec 2.2 "The range of possible outcomes in a series of estimates. For recoverable resource assessments, the range of uncertainty reflects a reasonable range of estimated potentially recoverable quantities for an individual accumulation or a project."

226 Petroleum Resources Management System 2007 Sec 2.2 "Key uncertainties affecting in-place volumes include:
- Reservoir geometry and trap limits that impact gross rock volume.
- Geological characteristics that define pore volume and permeability distribution.
- Elevation of fluid contacts.
- Combinations of reservoir quality, fluid types, and contacts that control fluid saturations."

227 Auditing Standards for Reserves, SPE, June 2001 2.3 "A subsurface rock formation containing an individual and separate natural accumulation of moveable petroleum that is confined by impermeable rocks/formations and is characterized by a single-pressure system."

The SPE gives some guidance to estimate reserves by the volumetric method.[228] The volumetric method involves using maps derived from well data and seismic data. Structure maps are used to define the surface of the reservoir and help identify reservoir limits such as gas oil contacts or water contacts. Isopachous maps, maps of equal thickness, are used with structure maps to estimate the reservoir's gross rock volume. The gross rock volume is then netted to a gross hydrocarbon volume by applying porosity and water saturation values estimated from log and core data. These are referred to as hydrocarbons in place. A recovery factor is applied to the in place hydrocarbon volume to arrive at the net volume that can be produced, the estimated ultimate recovery (EUR).[229] Remaining reserves (RRR) are the reserves available at a certain date after cumulative production is subtracted from the estimated ultimate recovery.

Reservoir Volume

The SEC defines a reservoir as "A porous and permeable underground formation containing a natural accumulation of producible oil and/or gas that is confined by impermeable rock or water barriers and is individual and separate from other reservoirs." The SPE defines a reservoir as "A subsurface rock formation containing an individual and separate natural accumulation of moveable petroleum that is confined by impermeable rocks/formations and is characterized by a single-pressure system." The major difference between the two definitions is the SPE adds the comment of being characterized by a single pressure system.

---

228 Auditing Standards for Reserves, SPE, June 2001 5.4 "Estimating reserves in accordance with the volumetric method involves estimation of petroleum in place based upon review and analysis of such documents and information as (i) ownership and development maps; (ii) geological maps and models; (iii) openhole and cased-hole well logs and formation tests; (iv) relevant reservoir, fluid, and core data; (v) relevant seismic data and interpretations; and (v) information regarding the existing and planned completion of oil and gas wells and any production performance thereof. An appropriately estimated recovery efficiency is applied to the resulting oil and gas in place quantities in order to derive estimated original reserves.

229 SPE, Petroleum Resources Management System 2007 Sec1.1 "Those quantities of petroleum which are estimated, on a given date, to be potentially recoverable from an accumulation, plus those quantities already produced therefrom."

Reservoir volume is a function of reservoir area and net pay. Net pay is generally considered to be that part of the reservoir expected to produce hydrocarbons at economic rates. The term gross formation interval or gross thickness is used to describe the total formation from top to bottom. Net thickness refers to the amount of the reservoir having porosity and water saturation values better than the cut off values. It describes the amount of the total reservoir having producible hydrocarbons. The more well control there is available and the better the understanding of the depositional environment, the better the estimate of net reservoir volume.

Often sands are made up of more than one member. This can lead to the question of how much vertical continuity is present. Are the members connected, or are they separate entities. One method to confirm vertical continuity is consistency of fluid contacts across the different sand members. Another indication of vertical continuity is depletion of one member by production from another. One can estimate a larger volume than exists if continuity is assumed where it does not exist.

The second factor needed to estimate reservoir volume is the reservoir area. This is determined by mapping on the top of the reservoir i.e. generally top of porosity, to the known or inferred reservoir limit. The reservoir limit may be a water contact, lowest known hydrocarbon, stratigraphic change, or fault. Seismic can be used to help define the structure, along with well control. One should be careful if seismic is used to define the structure that the event mapped on is the actual top of the reservoir and not a marker above it. One should also insure the seismic map is a depth map and not a time map.

The lowest known hydrocarbon is the lowest point of economic production. This cutoff can be determined based on tests or by analogy with areas which have been tested. The lowest known hydrocarbon contains both the presence of hydrocarbons and the ability to produce those hydrocarbons at a commercial rate. This is a concern in reservoirs which have a fining downward sand. The lowest known hydrocarbon should be picked at the point where the formation will no longer produce at economic rates.

The highest known hydrocarbon limit should also be considered in mapping the reservoir area. Depending on the type of drive mechanism, the area above the highest well may or may not be recoverable and should be treated accordingly in assigning mapped areas.

In general, the reservoir area should be delineated by wells and above the water contact or lowest known hydrocarbon limit and below the highest know hydrocarbon limit. The portion of the total mapped area actually used to estimate reserves should fit the estimated drainage areas for the drilled wells and any undeveloped locations.

Mapping

The uncertainty of in-place calculations is a function of reservoir complexity. In general, the simpler the structure and stratigraphy, the less uncertain the estimate is. For example, an anticline with no faulting and clean blanket sands is easy to map. A stratigraphically complex, highly faulted reservoir is difficult to map and the volumetric estimate is less certain.

Maps should be adjusted after each well is drilled. After the initial well has been drilled on a prospect, it should be integrated into the original seismic structure map along and any needed changes made. Seismic parameters should be adjusted to accommodate the data obtained from the well. Fluid contacts should be added to the structure maps. Gross and net sand isopach maps should be constructed and the fluid contacts added to them. The last maps to be constructed are the net hydrocarbon maps. After production has begun, additional maps may be needed to track the movement of the water contact and verify the in-place estimates and recovery factor.

There are a number of things to consider when constructing maps for a volumetric estimate.[230] Is the seismic structure map on the reservoir of interest? It should not be on a reservoir significantly higher or lower than the reservoir of

230 SPE 91069, Harrell, Hodgin and Wagenhofer, Oil and Gas Reserves Estimates: Recurring Mistakes and Errors

interest. Has the seismic time maps been converted to depth? The seismic map should be converted to a time map based on a velocity survey from a nearby well. If more than one survey is available, a velocity depth relationship can be calculated and mapped. Have seismic faults been confirmed by subsurface data or production anomalies?

Is the structure map drawn on the top of the effective reservoir? The structure map should not be on a marker above the reservoir. If the top of the reservoir is not porous, the structure map needs to be on the top of porosity and not the top of sand. If the top of reservoir map is not on the top of porosity, the size of the reservoir will be too large; the reservoir limits will be mapped too far from the well. Are the upper and lower limits based on well control? The SEC requires the lower limits to be based on wireline logs or tests. The SPE allows the use of MDT data and seismic data to estimate the water contact under certain conditions. The reservoir limits should not be based on seismic without confirming well control.[231]

Has reasonable judgment been used to place the zero isopach contour? The zero isopach should be place based on a projection of the existing thinning rate. Has the sand been wedged into a stratigraphic barrier? The sand should be wedged into the barrier unless the geologic model would suggest otherwise, for instance a shale filled channel down cutting into the reservoir. Has the correct net to gross been used for a sand coming out of water? Consideration should be given the concentration of sand in the different members. The net to gross may not be consistent over the vertical

231 Guidelines for the Evaluation of Petroleum Reserves and Resources, SPE "When a known gas accumulation is being appraised, it is reasonable that seismic flat spots and/or bright spots can be used as definitive geological data to classify gas as proved reserves when the following conditions are met:

- The flat spot and/or bright spot is clearly visible in the 3D seismic data.
- The spatial mapping of the flat spot and/or downdip edge of the bright spot fits a structural contour, which usually will be the spill point of the reservoir.
- A well penetrates the GWC in one fault block of the reservoir, so logs, pressure data, and test data provide a direct and unambiguous tie between the GWC in the well and the seismic flat spot and/or downdip edge of the bright spot; i.e., the borehole proves that there is producible gas, not residual gas, down to the seismic indicators of the GWC.
- A well in another fault block penetrates the reservoir updip from the GWC.
- This second well proves gas down to a lowest known depth, and pressure data show that this gas is in communication with the gas in the first fault block.
- The seismic flat spot and/or downdip edge of the bright spot in the second fault block lies below the lowest known gas in the second well and is spatially continuous with and at the same depth has these seismic indicators in the first fault block."

extent of the sand and the isopach should reflect the inconsistency.

Has correct net to gross been used for the wedge zone? The wedge should be contoured to reflect the sand as it rises above the water contact. If the net to gross is changing, the isopach contours should be spaced to accurately reflect this change. In the case of a changing net to gross, the contours should be spaced to reflect the average net to gross for the sand, not evenly spaced.

Has a common contact been used for a stratified reservoir without confirmation the members are part of the same reservoir? Doing so can result in an overly optimistic contact for the unit. Until there is confirmation the units are part of the same reservoir, the contact for each unit should be considered and the units mapped separately.

Reservoir Fluids

The SPE defines petroleum as "... as a naturally occurring mixture consisting of hydrocarbons in the gaseous, liquid, or solid phase. Petroleum may also contain non-hydrocarbon compounds, common examples of which are carbon dioxide, nitrogen, hydrogen sulfide, and sulfur. In rare cases, non-hydrocarbon content could be greater than 50%."[232] Hydrocarbons are defined as "...chemical compounds consisting wholly of hydrogen and carbon."[233]

The reservoir fluids can be in a liquid or gaseous phase. Oil is in a liquid phase both in the reservoir and at the surface. Natural gas can be solution gas which is dissolved in the oil at reservoir conditions. Natural gas can be associated gas, or gas cap gas, which is in a gaseous state both in the reservoir and at the surface. Non-associated gas refers to the gaseous phase only, without an associated oil rim. Gas condensate is dissolved in the gas in the reservoir and is a liquid phase at the surface. Natural gas liquids, sometimes referred to as plant products, are liquids which are extracted from the gas at the surface by processing plants.

232 SPE, Petroleum Resources Management System 2007 Sec1.0.

233 SPE, Petroleum Resources Management System 2007 Sec1.1.

There are by-products associated with oil and gas reservoirs such as sulfur, helium, carbon dioxide, hydrogen sulfide. Although these products are produced from the reservoir, they are not included in reserve volumes.

Crude oil is a widespread liquid phase hydrocarbon. It is described by its API gravity, its viscosity, its bubble point and its gas content or GOR. It is expressed in barrels, cubic meters, and tons. A barrel is 42 U.S. gallons. A cubic meter is 6.2898 barrels. A ton varies depending on the gravity of the oil and there is generally about 7 barrels per ton. Crude oil generally ranges from an API gravity of 8 (heavy oil) to about 45. API gravities higher than 45 are generally associated with gas condensate. Fresh water has an API gravity of 10.

Natural gas is measured in cubic feet which is dependent on the pressure base used where the gas is produced. Pressure bases range from about 14.65 to 15.025. Natural gas is commonly referred to as dry gas which implies very little condensate, or wet gas which has a higher condensate yield. A third type is the retrograde gas which has a variable condensate yield. As the reservoir pressure drops, the liquid will start to drop out in the reservoir. Gas is also referred to as sweet, which implies no hydrogen sulfide or carbon dioxide and sour gas which does.

The sales volumes of natural gas must be adjusted for the removal of by-products and for condensate. Both of these additions result in a lower sales volume. The difference between the sales volume and the produced volume are accounted for by a shrinkage factor, the value of which will depend on the amount of each by-product and the amount of condensate produced. For condensate, the rule of thumb is about 1% shrinkage for each 10 bbl/mmcf of condensate.

Porosity

Porosity is the open space in the reservoir matrix, for instance, the voids between the sand grains in a sandstone reservoir. It can also include vugs in carbonate reservoirs or fracture systems in some reservoirs. It is commonly filled with fluid, either water or hydrocarbons. In some cases, the pore space is filled with secondary cement and part, or all, of the initial reservoir volume lost.

Conversely, porosity can be enhanced by the digenetic replacement of certain minerals by others.

Porosity is commonly calculated from wireline logs or measured directly from cores. The cores can be whole cores or sidewall cores. If cores are available, the critical water saturation and permeability can also be determined.

Wireline logs, however, are the most common method for determining porosity, and various kinds of logs are used. The decision to chose a particular type of log will depend on the reservoir geology, the geologic area, operator preference, cost, hole conditions, and reservoir temperature and pressure. Most geologists prefer the Neutron-Density log. It generally gives more reliable readings and requires fewer assumptions than other types of logs. Other wireline logs used include the sonic log, the Microlog, and some cased hole logs. If cores are available, core porosity should be plotted against log porosity to insure the correct assumptions are being used and appropriate corrections are being applied.

The whole core porosity is the usual standard the log porosity is calibrated against. If the reservoir contains vugs or fractures, neither core nor may log porosities be reliable. If the reservoir is friable sandstone, the core may yield porosities which are too high.[234]

The porosity calculated from logs is apparent porosity and must be adjusted for certain reservoir conditions and fluids. The corrected porosity is called the effective porosity. Readings from wireline logs must be adjusted for the amount of shale in the formation of interest and the kind of fluid or gas. The readings must also be corrected if the matrix assumed for the log run is different than that of the reservoir being evaluated. There are various formulas available in the literature to make these corrections. Fracture porosity is determined separately from matrix porosity.

The average reservoir porosity used in hydrocarbon in place calculations must be

234 Cronquist, C., "Estimation and Classification of Reserves of Crude Oil, Natural Gas, and Condensate", p.53.

adjusted to exclude parts of the reservoir that are not productive, that is, parts of the reservoir where porosity is so low there is not adequate permeability for fluid flow.

There is usually a linear correlation between porosity and permeability. If cores are available, the relationship can be plotted and a minimum porosity value determined based on the minimum permeability for fluid flow. If cores are not available, then analogy to other wells in the reservoir can be used to make this assumption.

The National Standard of PRC defines effective porosity as the interconnected porosity in the rock.[235] It should be based on core analysis, but if logs are used, there should be good agreement with the cores – within 1-1.5%.[236] The value to be used is the value at the formation depth and not the surface. In carbonate or fractured reservoirs, the matrix and fracture or vugs should be determined separately. Porosity in vugs, fractures, and gravels should be based on large diameter cores. In unconsolidated sandstones, the "frozen analysis" method should be used.[237]

Water Saturation

The SEC does not say how water saturation should be determined, nor does the SPE/WPC. The particular method used, however, should be reasonable for the reservoir in question and accurately portray the relative fluid content. The assumptions made should be documented with the reasons a particular value was used.

Pore space in reservoirs is filled with fluid. The hydrocarbon saturation is the amount of that pore space that is hydrocarbon filled. The remainder of the pore space is water and this is referred to as water saturation. The water saturation

235 The National Standard of P.R.C., Petroleum Reserve Standard.

236 Id.

237 Id.

at which a reservoir will flow water instead of hydrocarbons is referred to as the critical water saturation. If the water saturation in the formation is higher than the critical water saturation, the reservoir will flow water instead of hydrocarbons. The critical water saturation can be estimated from cores or relative permeability curves.

The water saturation range which will allow for water free hydrocarbon production is generally from 35% to 65%. The critical water saturation for a particular reservoir will depend on a number of factors. It is dependent on grain size, with smaller grain sizes being more likely to produce water free at higher calculated water saturations. It will depend on the type of water in the reservoir, whether it is bound water or movable water. If there is a larger amount of bound water than the reservoir can flow hydrocarbons at a higher calculated water saturation. If there is not a significant transition zone, the average water saturation may approach the irreducible water saturation.[238]

The water saturation cutoff used in reserve calculations can be determined in various ways. The most direct way is by tests. It can also be estimated from core analysis or by analogy to other wells in the area that produce from the formation.

There are a number of formulas in the literature which can be used to calculate water saturation. The simplest is the Archie Equation. The Archie Equation is typically used in clean sandstones or carbonates. If there is shale in the matrix, the Simendoux Equation can be used to account for the effect shale has on the apparent saturation calculated. There are also Dual Water Models and other more complex algorithms available for water saturation calculations. The important thing is to be able to document the assumptions made and the reason a particular method was chosen for the calculation.

The National Standard of PRC says the oil saturation is to be determined based on wireline logs. It is defined as the percentage oil volume in the effective

---

238 Cronquist, C., "Estimation and Classification of Reserves of Crude Oil, Natural Gas, and Condensate", p.51.

porosity at original reservoir conditions.[239] If the field size is over 100 million tons in place, it must be verified and a core taken for comparison. Maps based on oil saturation and height above the oil water contact relationship should be constructed. In a fractured reservoir, the oil saturation in the matrix and the fractures should be determined separately. The saturation used in estimating proved reserves can be based on logs, capillary pressures or analogy to a nearby field.

Pressure Base

The SEC notes that in the instructions for the Department of Energy's Form EIA-28, gas reserves are to be reported at 14.73 psia and 60 degrees F. They do not object to other pressure bases, however, if they are identified in the filing.[240] Typically, the pressure base where the reserves are located is used and noted in the report.

The PRC states that reserves should be reported at standard conditions of 20 degrees Centigrade and 0.101 MPa.[241]

Reservoir Temperature and Pressure

Reservoir temperature can be obtained in from the log headings, estimated from the regional temperature gradient, or taken from other downhole readings such as pressure tests.

Reservoir pressures can be measured, estimated based on the mud weight used to drill the well, or estimated from the regional pressure gradient. The

239 The National Standard of P.R.C., Petroleum Reserve Standard.

240 Topic 12 of Accounting Series Release No. 257 of the Staff Accounting Bulletins "Question 4: What pressure base should be used for reporting gas and production, 14.73 psia or the pressure base specified by the state? Interpretive Response: The reporting instructions to the Department of Energy's Form EIA-28 specify that natural gas reserves are to be reported at 14.73 psia and 60 degrees F. There is no pressure base specified in Regulation S-X or S-K. At the present time the staff will not object to natural gas reserves and production data calculated at other pressure bases, if such other pressure bases are identified in the filing."

241 GAO Ruiqi, LU Minggang, ZHA Quanheng, XIAO Deming, HU Yundong, China Petroleum Resources/ Reserves Classification.

initial pressure should be compared to the regional gradient, especially in new reservoirs. If the pressure is subnormal, than it is important to know whether the low pressure is geologic or the result of partial depletion from adjacent production.[242]

Reservoir temperature and pressure are estimated at the midpoint of the reservoir. If there is a significant gas cap with an oil reservoir, it may be appropriate to estimate the temperature and pressure at the gas-oil contact.[243]

Formation Volume Factor

The formation volume factor determined from differential liberation should be adjusted to reflect field separator conditions.[244] If the formation volume factor is determined from empirical correlations, the estimated bubble point pressure should be checked to insure it is less than or equal to the actual reservoir pressure.[245]

Cutoffs

Cutoffs are used to determine what portion of the gross sand is net pay. Net pay is generally defined as that part of the reservoir which will produce hydrocarbons at commercial rates. To indentify net pay, we determine the values for porosity, water saturation and shale volume outside of which the particular reservoir will not produce at economic rates. Cutoffs eliminate hydrocarbons which are unlikely to be recovered from the reserve estimate.

Arbitrary cutoffs should not be used. Some support is needed such as tests, cores, or analogy. The assumptions used to estimate the cutoffs should be documented.

---

242 Cronquist, C., "Estimation and Classification of Reserves of Crude Oil, Natural Gas, and Condensate", p.54.

243 Id.

244 Cronquist, C., "Estimation and Classification of Reserves of Crude Oil, Natural Gas, and Condensate", p.171.

245 Id.

With clean, homogenous sand, cutoffs are not generally needed, and bed boundaries are the only concern. If the reservoir is homogenous, one test may be sufficient to establish cutoff values. If the reservoir is not homogenous, it may not be possible to determine where the flow is coming from if the members are not tested individually. Multiple tests may be required.

Porosity cutoffs should be at the minimum porosity with sufficient permeability for commercial flow. Porosity cutoffs can be determined by porosity vs. permeability plots based on core data. Drill stem tests and production tests can be used. Analogy to other producing wells in the formation is another method available. The actual porosity cutoff can range from about 6% in tight reservoirs to about 18% in clean, porous sands. The ranges above are general and should not be taken as absolute. The porosity cutoff depends to a large degree on grain size.

As shale volume increases, the effective permeability decreases. One needs to cautious if the shale volume is over 40%. Need to calibrate the calculation method to tests or cores to determine the actual shale volume. Different shale volumes are implied using different formulas.

Water saturation cutoffs are generally based on the critical water saturation. It can be determined using cores, tests, or analogy. The water saturation cutoff can range from about 40% in clean, coarse grained sands to about 65 – 70% in shaly, fine grained sands. Bound water will show up in the water saturation calculation, but the water will not flow.

Questionable zones should be tested individually to determine if they will contribute to economic production from the reservoir. The intent is to have a reasonable estimate of the amount of hydrocarbons which can be recovered from the reservoir. The net pay should be questioned if the porosity cutoff is too low, i.e. less than 6%, if the water saturation cutoff is too high, i.e. greater than 65%, or the shale content is too high i.e. greater than 40%. One should have supporting documentation for the cutoffs used, especially if outside the parameters noted above.

Reservoir Limits

The downdip reservoir limits are based on the intersection of the water contact or lowest known hydrocarbon on the top of sand structure map. This becomes the zero line for the net hydrocarbon isopach. The intersection of the water contact on the base of sand structure map is the inner limit of the water. The net pay contours above this limit are controlled by net sand contours from the net sand map. The area between the lower water limit and the inner water limit is known as the wedge zone. Net pay contours in this area are controlled by the vertical distribution of net sand and the structural gain above the downdip limit. The sand/shale distribution must be considered in the wedge zone to insure the proper gain in net pay.

The steps in mapping a reservoir include first making structure maps on the top and base of sand. Next a net sand isopach map is constructed. The last step is to construct a net pay isopach as discussed above.

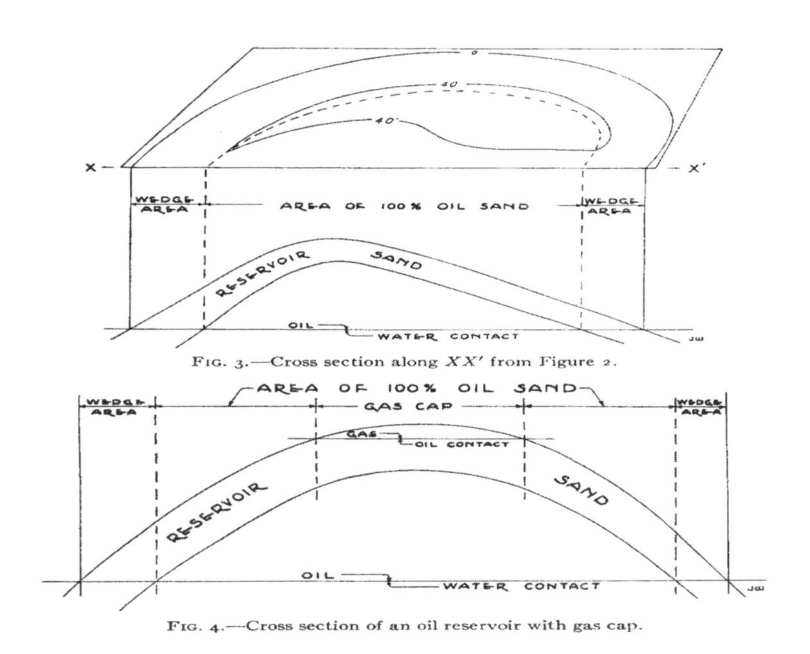

FIG. 3.—Cross section along XX' from Figure 2.

FIG. 4.—Cross section of an oil reservoir with gas cap.

• WHARTON, JAY B., Isopachous maps of sand reservoirs: Bulletin of the American Association of Petroleum Geologists, v. 32, No. 7 (July, 1948),

The SEC required a reservoir limit based on wireline logs, and restricted the limit to the lowest known hydrocarbon, in the absence of a water contact in the well prior to the December 31, 2008 revision. The SEC also did not allow the use of MDT data or seismic data to estimate the reservoir limit. With the December 31, 2008 revision to the reserve guidelines, the SEC has said new reliable technologies can be used to establish the limits.[246]

Company management does not always want to drill to find the water contact and thus the reservoir limits. What other ways are available to determine reservoir limits?

The water contact can be determined from tests. The SPE/WPC allows the use of MDT data to estimate a water contact below what can be seen on the wireline logs.

The SPE/WPC also allows the use of seismic to estimate the water contact under certain conditions. The SPE/WPC will allow the water contact to be estimated with seismic for a fault block adjacent to a fault block with a known water level.

The water contact is generally thought of as the free water level. Some reservoirs may have a transition zone and it may not be appropriate to consider the entire transition zone as recoverable. Transition zones are usually seen in low permeability reservoirs, but can also be associated with heavy oil reservoirs. The water contact should be placed where the calculated water saturation is equal to the critical water saturation.

China allows the water contact to be estimated if there is no contact in the well. It is placed at a point one half the distances between the lowest known hydrocarbon and the top of the next wet sand. The China definitions also allow the use of pressure data to estimate a water contact below the lowest known hydrocarbon.

246 Securities and Exchange Commission, Modernization of Oil and Gas Reporting, December 31, 2008; p.26.

Reservoir Area

For proved reserves, the reservoir area is considered to be the area delineated by drilling.

Recovery Factor

A recovery factor is applied to an in place volume of hydrocarbons to determine the volume which can actually be recovered. This recovery factor is a function of the reservoir fluid and the fluid properties, rock properties, drive mechanism and development practices. As additional data becomes available in the form of production history and pressures, the recovery factor may need to be modified.

The SPE says the portion of the in-place volume that can be recovered under a certain defined set of conditions must be estimated based on analogy and/or simulation studies using available information.[247] The estimated recovery efficiency is applied to the in place volume to derive the original reserves.[248]

The SPE allows the use of different recovery factors for additional probable and possible reserves. The recovery factors used for probable and possible reserves may be higher than for proved reserves due to lack of available data. For instance, a gas reservoir may be assigned a proved recovery factor based on the assumption of a water drive and a probable and possible recovery factors based on the assumption of a partial water drive or depletion drive recovery. Another factor may be the uncertainty in the availability of compression and so a higher recovery factor is used for additional probable reserves based on the assumption compression may be available at some point. If a good analogy does not exist,

---

247 SPE Petroleum Resources Management System, March 2007 – 4.1.2 "Given estimates of the in-place petroleum, that portion that can be recovered by a defined set of wells and operating conditions must then be estimated based on analog field performance and/or simulation studies using available reservoir information. Key assumptions must be made regarding reservoir drive mechanisms."

248 SPE Petroleum Resources Management System, March 2007 – 5.4 "An appropriately estimated recovery efficiency is applied to the resulting oil and gas in place quantities in order to derive estimated original reserves. The unmodified term "reserves" is applicable to remaining quantities of petroleum, net of cumulative production, at any effective reporting date. The estimated recovery efficiency may also vary as a function of the appropriate reserves classification."

then the proved recovery factor may be increased as additional probable or possible reserves. In overpressure reservoirs, the formation compressibility is not always known with any certainty. The more pessimistic option should be used for proved and additional probable reserves assigned based on a more favorable assumption.

There are formulas available in the literature to calculate recovery factors. These formulas apply to ideal conditions and do not account for reservoir heterogeneities or operational differences. Even the position of the well in the reservoir or the well spacing can affect the recovery factor. If the well is located in a downdip position or in one corner of the reservoir, the recovery factor will be less than the calculated. The calculated recovery factor should be adjusted to account for factors such as spacing and well position in the reservoir. Often this requires engineering judgment.

In order to accurately determine the recovery factor, one must determine the type of drive mechanism. This can be inferred by using analogy or inferred from the geologic setting. In using the geologic setting, one should determine if the reservoir is open to water or is a closed system. A closed system would imply a depletion drive reservoir and if it is open to water, one would expect some water support. If the reservoir is stratigraphic, one would expect a depletion drive reservoir. If the reservoir is bounded by faults or is basin centered gas, then would also expect a depletion drive reservoir. One should also consider if the reservoir is overpressured, as this will affect the recovery factor. Generally, recovery factors in overpressured gas reservoirs are lower than depletion reservoirs.

Reservoir heterogeneity also affects the recovery factor, resulting in the nonuniform recovery of hydrocarbons. These include shale stringers which may trap some hydrocarbons by not allowing hydrocarbon flow. Fractures affect the recovery factor. Lateral reservoir discontinuities affect the flow of hydrocarbons. Finally, vugular or fracture controlled porosity in carbonates affects the recovery factor.

In oil reservoirs, one would expect a recovery factor in the range of 15% to 30% for solution gas drive reservoirs and 35% to 60% for water drive reservoirs. In limestones, one would expect a recovery factor of about 20% for solution drive reservoirs and about 45% for water drive reservoirs.

The recovery factor in a depletion drive gas reservoir depends on the initial reservoir pressure and the abandonment pressure of the reservoir. Generally, the range of recovery factors is 50% to 85%. If the gas reservoir is a water drive reservoir, the recovery factor depends on the strength of the water drive. Generally, it ranges from 40% to 75%.

Another important factor in the estimation of the recovery factor is the development plan. The recovery factor used to estimate the original reserves must consider how the reservoir will be developed and produced.[249] One should consider the well spacing and where the wells are located within the reservoir. Large drainage areas imply a lower recovery factor. If the well is located near the water contact or in one corner of the reservoir, one should expect a lower recovery factor. One should consider how the wells are to be completed and how they are to be perforated. If more than one zone is open, this may affect the recovery factor of each individual zone. If the well is perforated near the water contact, the recovery factor may be affected. The pipeline pressure and availability of compression should be considered. The economics should also be reviewed, such as the cost to operate the well and the field.

One should always document the assumptions used in estimating the recovery factor and the reasons they were used. One should note why they assumed a particular drive mechanism. One should also consider whether the recovery factor estimated is reasonable when compared to other production in the field or in adjacent fields.

At times the volumetric estimate may be higher than the dynamic estimates

249 SPE Petroleum Resources Management System, March 2007 – 4.1.2 "The estimates of recoverable quantities must reflect uncertainties not only in the petroleum inplace but also in the recovery efficiency of the development project(s) applied to the specific reservoir being studied."

made later in the life of the field. Some operators want to book undeveloped based on the difference. A difference between the static and dynamic methods may in fact indicate additional reserves to be recovered. It may point to areas see by performance data and not mapped, or the static method may infer the current producers are not adequately draining the reservoir. Care should be taken, however, to insure the difference is not due to an unrealistically high recovery factor. Before reserves are booked based on a difference between the volumetric and dynamic results, additional data should be available to confirm the volumes.

Data Requirements

One should also consider the data needed to do a volumetric estimate. In general, we need data on the rock properties, fluid properties and reservoir properties. We need information about the porosity, permeability and water saturation. We need information about viscosity, gor/yield, the density of the fluid, what contaminants are present, and the BTU value of the gas. We need to determine the reservoir volume from structure and isopach maps. From the development plan, we can determine the number of additional wells to be drilled, when they will be drilled and their locations.

Data comes from three basic sources. It comes from wireline logs, cores and tests and fluid samples. Minor sources of data include cuttings and mudlogs.

Logs and cores are the most sources of data for a volumetric estimate. They provide data on rock properties and mapping parameters. They are the most important source of data for a volumetric estimate of reserves.

Logs provide information about porosity, litholgy, water saturation, reservoir limits, correlations and fluid contacts. Logs are considered by many to be one of the most important sources of data.

Cores provide information about the porosity, lithology, critical water saturation, type of reservoir fluid, permeability, and fractures and vugs. Primarily, cores provide information about the rock properties.

Mud logs and test provide information about reservoir pressure and temperature, flow capacity and permeability, reservoir limits, lithology, fluid contacts, and fluid properties.

Wireline logs are generally run to estimate certain rock and reservoir properties. The logs usually run include the SP, the gamma ray, the sonic, the resistivity, the density and the neutron logs. Other logs may be run as needed or all of the logs mentions above may not be run depending on the data required to be collected and the constraints placed by the operator.

The SP and gamma ray logs are used for correlations and to help identify potential zones of interest. The gamma ray log is also used to calculate the shale volume of the formation.

Information about the reservoir porosity is obtained from the density, neutron and sonic logs. Usually, the density and neutron are run together and provide a relatively reliable source of porosity data. Sometimes due to environmental or operational issues only the sonic log may be run. The density and neutron logs are often run together and provide information about the lithology, porosity and fluid in the formation. If the wrong matrix is used, however, the calculated porosity will be in error. The density log measures total porosity. The readings should be checked against known lithologies to insure the readings are correct. The sonic log can be used to identify secondary porosity and fractures.

Water saturation calculations require porosity and resistivity logs. In older wells, the electric log was used. In the newer wells, the induction log is used. It has the advantage of deeper readings, better invasion profile and better bed resolution.

Two basic types of cores are taken, whole cores and sidewall cores. Whole core provide the most reliable information. They may not always be taken because of the cost, time and risk involved. They can be used to determine porosity and permeability and to calibrate wireline log readings. They can be used to visually check for fractures or vugs. They can also be used to determine lithology and deposition environments. Sidewall cores are a good alternative to whole cores.

They are often used to determine porosity and permeability. Care should be exercised however, since they may read high in low porosity formations, due to induced fractures during the coring, or too low in high porosity, unconsolidated formations due to compaction during the coring process. Sidewall cores can be used to determine lithology and matrix density. Sidewall core can also be useful on determining fluid contacts.

Mudlogs provide information on correlations, lithology, and reservoir fluid.

RFT/MDTs can be used to estimate reservoir temperature and pressure. They have also been used to determine reservoir contacts. While the SPE allow the use of this data to determine the fluid contacts, the SEC has been slow to accept this as a reliable means of estimating fluid contacts. DST's can provide data on reservoir temperature and pressure. It can be used to estimate permeability and flow characteristics. It can be used to estimate reservoir boundaries. It can be used to obtain a fluid sample to estimate fluid properties.

Comment Letters

Comment letters sent to a company by the SEC may include a series of questions designed to ensure compliance with SEC reporting requirements. The initial set of questions may be followed by a smaller set the next round and so on until the SEC is satisfied reserves are being properly reported. The end result may be as simple as no more letters or the SEC may require de-booking or a restatement of reserves. Examples of SEC comment letters have been included in Appendix F. Questions may relate to whether or not reserves have been booked below the low known hydrocarbon without being supported by actual production, whether year end prices were used, whether any undeveloped locations are booked more than one location from a commercial well, or whether reserves were booked based on information other than a test. The SEC has inquired whether or not bonuses are tied to reserve increases. The SEC has also asked whether or not an outside reserve auditor has been used, who has been used, how long they have been used, who has the authority to hire them, who do they report to, and how much they were paid. They have asked producers to reconcile proved undeveloped reserves and justify the bookings by comparing the pre-drill to the post drill reserves. They have asked for exhibits and explanation of the differences as well as corporate policy for reducing the difference. They have asked for similar information for large reserve revisions. They have asked for an explanation of differences between projected and actual capital expenditures. They have asked explanations for wells projected to be drilled that were not drilled and if they are still carried as proved, why.

Producers have been asked to comment on the internal controls in place and their effectiveness in controlling reserve revisions. They may ask who has the final say on reserve bookings.

Comment letters are now to be made public by the SEC through their website at www.sec.com.[250] However, the SEC may still hold some information confidential

250 SEC Press Release June 24, 2004 & May 9, 2005

if requested by the company under Rule 83 (17CFR.83). In this case, two copies of the comment letter response will be requested by the SEC. One, marked as confidential, which includes the information the filer is requesting be held confidential, and another, without the confidential information-redacted. The redacted version is the only copy the SEC will make public.

For help in finding oil and gas comment letters, one can go to www.sec.gov. answers/edgarletters.htm.

Reserve Estimates

General

There are two reserve estimate methods, deterministic and probabilistic. The deterministic method involves estimating reserves based on the selection of a single value for each parameter. It is the best estimate based on the existing geological, engineering and financial data. The probabilistic method uses a range of possible values for each parameter and their probabilities. The SEC does not address probabilistic reserves, but the SPE/WPC require a 90% certainty of the reserves for them to be classified as proved.

The two most important requirements for quality reserve estimates are the data and the evaluator. There are four main methods to estimate reserves. They include the volumetric method, which is used if the there is no production data or the production data is not definitive; the performance method, using the projection of past production to estimate future production; there are the material balance and simulation methods as more performance and pressure history becomes available; and there is analogy to other reservoirs if there is not enough geologic or performance data to base an estimate on.[251] Various methods should be used and the results compared to determine the best answer.

The Petroleum Reserve Standard states reserves should be determined based on reservoir maturity. In the early stages, the volumetric method should be used,

251 Auditing Standards for Reserves, SPE, June 2001 Sec 5.3 "The acceptable methods for estimating reserves include (i) the volumetric method; (ii) evaluation of the performance history, which evaluation may include an analysis and projection of producing ranges, reservoir pressures, oil-water ratios, gas-oil ratios and gas-liquid ratios; (iii) development of a mathematical model through consideration of material balance and computer simulation techniques; (iv) analogy to other reservoirs if geographic location, formation characteristics or similar factors render such analogy appropriate. In estimating reserves, Reserve Estimators should utilize the particular methods, and the number of methods, which in their professional judgment are most appropriate given (i) the geographic location, formation characteristics and nature of the property or group of properties with respect to which reserves are being estimated; (ii) the amount and quality of available data; and (iii) the significance of such property or group of properties in relation to the oil and gas properties with respect to which reserves are being estimated."

and then later in the reservoir's life, the performance method should be used to check the volumetric reserves. It also states advanced technology should be used and all basic data accurately gathered. The Petroleum Reserve Standard lists a table of standard cutoffs of economic flow for wells at different depths. It is noted that these are general conditions only and they can be adjusted based on prices, location, and technology.

Data

Reserve estimates are dependent on the amount and quality of data available to the reserve evaluator. There is a direct correlation between the amount and quality of data and the reliability of the reserve estimate. Without good quality data the estimate will be weak at best. Reserves may be understated without adequate data, as current SEC guidance requires the evaluator to error on the side of conservatism. As the SEC says, proved reserves are more likely to go up with additional data than go down.

Reserve Evaluator

The second part of the reserve equation is the reserve evaluator. The evaluator, whether he is a geologist, geophysicist, or engineer, must be qualified. The SPE suggests he has at least three years experience with at least one year in reserve work, he has a degree in engineering or geology, and he is registered or certified. The evaluator should have adequate training in the best methods of reserve estimation, and he must understand the reserve definitions being applied.[252] The

252 Auditing Standards for Reserves, SPE, June 2001 Sec 3.2 "A Reserve Estimator shall be considered professionally qualified in such capacity if he or she has sufficient educational background, professional training and professional experience to enable him or her to exercise prudent professional judgment and to be in responsible charge in connection with the estimating of reserves and other Reserve Information. The determination of whether a Reserve Estimator is professionally qualified should be made on an individual-by-individual basis. A Reserve Estimator would normally be considered to be qualified if he or she (i) has a minimum of three years' practical experience in petroleum engineering or petroleum production geology, with at least one year of such experience being in the estimation and evaluation of Reserve Information; and (ii) either (A) has obtained, from a college or university of recognized stature, a bachelor's or advanced degree in petroleum engineering, geology or other discipline of engineering or physical science or (B) has received, and is maintaining in good standing, a registered or certified professional engineer's license or a registered or certified professional geologist's license, or the equivalent thereof, from an appropriate governmental authority or professional organization."

reserve evaluator must also have ethics.

Whether the reserve evaluator is employed by the company itself or is a third party consultant hired by the company, he should be independent. The reserve evaluator should base his conclusions on data and not on the company's internal politics or management bonuses. To avoid questions, an ethical relationship between the company and the evaluator should be maintained.

Sarbanes-Oxley implies the evaluator's estimate must be independent and the reserve process transparent. Corporate governance requires the reserve evaluator to have access to the data and the leeway to interpret the data to provide a competent opinion on reserves. He must be able to report the results without fear of repercussions. The primary objective of Sarbanes-Oxley is to provide transparency and reliability in corporate governance and for oil and gas companies; this would imply transparency and reliability in reserve reporting. The reserve evaluator should provide a report to the company explaining the reserves in layman's term and any qualifications to the reserves.

Deterministic

The deterministic method for estimating reserves involves making the estimate of reserves based on known geological, engineering and economic data using the best estimate for each parameter.

In its December 31, 2008 guidelines, the SEC defines the deterministic method as an "... estimate based on a single value for each parameter (from the geoscience, engineering, or economic data) in the reserves calculation that is used in the reserves estimation procedure."[253]

The SEC favors the deterministic method for reserve determination. The standard used is "reasonable certainty" which the SEC says indicates a high degree of certainty. Although the personal computer has made probabilistic

253 Securities and Exchange Commission, Modernization of Oil and Gas Reporting, December 31, 2008; p.42 and 17 CFR 210.4-10(a)(5)

reserve estimates more useful, the SEC still sees many cases in which the median is used for reserves. The SEC feels this is not within the bounds of reasonable certainty.[254] They have said reasonable certainty implies a reserve volume that will more likely increase than decrease with time and additional data.[255]

The SPE/WPC states that if the deterministic method is used to estimate proved reserves there should be a high confidence the reserves will be recovered.[256] Probable reserves are defined as being more likely than not to be recovered.[257] Possible reserves are less likely to be recovered than probable reserves.[258]

Probabilistic

The probabilistic method of reserve estimates uses a range of estimates for each parameter and their probabilities of occurring.

In its December 31, 2008 guidelines, the SEC defines the probabilistic method as "... an estimate that is obtained when the full range of values that could reasonably occur from each unknown parameter (from the geoscience and engineering data) is used to generate a full range of possible outcomes and their

254 SEC Division of Corporation Finance: Frequently Requested Accounting and Financial Reporting Interpretations and Guidance, March 31, 2001, 3(i) "Probabilistic methods of reserve estimating have become more useful due to improved computing and more important because of its acceptance by professional organizations such as the SPE. The SEC staff feels that it would be premature to issue any confidence criteria at this time. The SPE has specified a 90% confidence level for the determination of proved reserves by probabilistic methods. Yet, many instances of past and current practice in deterministic methodology utilize a median or best estimate for proved reserves. Since the likelihood of a subsequent increase or positive revision to proved reserve estimates should be much greater than the likelihood of a decrease, we see an inconsistency that should be resolved. If probabilistic methods are used, the limiting criteria in the SEC definitions, such as LKH, are still in effect and shall be honored."

255 SEC Division of Corporation Finance: Frequently Requested Accounting and Financial Reporting Interpretations and Guidance, March 31, 2001, 3(a) "The concept of reasonable certainty implies that, as more technical data becomes available, a positive, or upward, revision is much more likely than a negative, or downward, revision."

256 SPE/WPC Reserve Definitions "If deterministic methods are used, the term reasonable certainty is intended to express a high degree of confidence that the quantities will be recovered."

257 SPE/WPC Reserve Definitions "Probable reserves are those unproved reserves which analysis of geological and engineering data suggests are more likely than not to be recoverable."

258 SPE/WPC Reserve Definitions "Possible reserves are those unproved reserves which analysis of geological and engineering data suggests are less likely to be recoverable than probable reserves."

associated probabilities of occurrence."[259]

The SEC does not disallow probabilistic reserve estimates. They indicate they will accept probabilistic reserves which are "professionally" prepared. They also state that if probabilistic methods are used for reserve determination, the limiting criteria, such as low known hydrocarbons, are still in effect.[260] The SEC requires a "straight forward" reconciliation of reserves, stating that probabilistic aggregation can result in larger reserve numbers due to the decrease in uncertainty, especially at the total company level.[261] The SEC may allow probabilistic aggregation at the field level.

Using the probabilistic method to determine reserves, the SPE/WPC says that a 90% probability of recovering the reserves is appropriate for proved reserves.[262] For the probabilistic method of reserve determination, the proved plus probable case should have at least a 50% probability of those reserves or more and possible should have at least a 10% chance or more of recovering those reserves.

Performance Methods

Performance methods include projections based on declining production rates or other reliable trends in pressure or production. Records are kept of production volumes of oil, gas and water, along with periodic measurements of well head and bottom hole pressures. Various types of graphs can be constructed using the

---

259 Securities and Exchange Commission, Modernization of Oil and Gas Reporting, December 31, 2008; p.42 and Rule 4-10(a)(19) [17 CFR 210.4-10(a)(19)].

260 SEC Division of Corporation Finance: Frequently Requested Accounting and Financial Reporting Interpretations and Guidance, March 31, 2001. 3(j) "The SEC staff feels that it would be premature to issue any confidence criteria at this time. The SPE has specified a 90% confidence level for the determination of proved reserves by probabilistic methods. Yet, many instances of past and current practice in deterministic methodology utilize a median or best estimate for proved reserves. Since the likelihood of a subsequent increase or positive revision to proved reserve estimates should be much greater than the likelihood of a decrease, we see an inconsistency that should be resolved. If probabilistic methods are used, the limiting criteria in the SEC definitions, such as LKH, are still in effect and shall be honored. Probabilistic aggregation of proved reserves can result in larger reserve estimates (due to the decrease in uncertainty of recovery) than simple addition would yield. We require a straight forward reconciliation of this for financial reporting purposes."

261 Id.

262 SPE/WPC Reserve Definitions "If probabilistic methods are used, there should be at least a 90% probability that the quantities actually recovered will equal or exceed the estimate."

data collected from the producing wells. A study of these historical production trends can then be used to estimate future production trends of the reservoir.

Reserves can be estimated using the rate of production decline vs. time.[263] Decline curve analysis can include extrapolation of historical production using exponential, hyperbolic, or harmonic declines. The production rate can also be plotted vs. cumulative production. Pressure over Z can be plotted vs. cumulative production, or pressure vs. time can be plotted to estimate remaining reserves. Water-oil ratios vs. cumulative production can be used. A plot of production rate times flowing tubing pressure vs. cumulative production can be useful to estimate ultimate reserves. The reserves estimated by performance methods are subject to an economic limit within the reserve report.

The p/z method is a material balance method used most effectively on a dry gas reservoir with no water drive. It can also be used on gas condensate reservoirs if the cumulative condensate is converted to a gas equivalent when plotted against p/z. It should be used on a reservoir basis. The OGIP(original gas in-place) should be checked at different points in time to insure it is not changing. An increasing OGIP would be indicative of a water drive reservoir and an invalid result. Low permeability reservoirs are also a problem for this method as reservoir pressures may not be built up enough to be meaningful. The p/z results should be compared to the volumetric estimate to further insure a valid answer. There should be a good match of historical data.

The production decline method using production rate vs. time is a frequently used method to estimate reserves. It is based on a decline in reservoir energy over time. If there is aquifer support, other factors should also be considered such as water cut and total fluid production. In addition to an estimate of remaining reserves, this method also provides a production forecast for the reserve estimate.

263 Auditing Standards for Reserves, SPE, June 2001 Sec 5.5. "For reservoirs with respect to which performance has disclosed reliable production trends, reserves may be estimated by analysis of performance histories and projections of such trends. These estimates may be primarily predicated on an analysis of the rates of decline in production and on appropriate consideration of other performance parameters such as reservoir pressures, oil-water ratios, gas-oil ratios and gas-liquid ratios."

When using a hyperbolic decline, a minimum decline rate should be used. It can be based on other wells in the reservoir which have a longer producing life or by analogy to production from the same formation in a nearby field.

Reservoir simulation can also be used as more data, especially pressure data becomes available. It is reliable if a good match can be made of predicted fluid production and pressure to the historical fluid production and pressures.

Unconventional Reservoirs

Unconventional reservoirs present challenges not observed with conventional reservoirs. There has been much discussion concerning the way unconventional performance estimates should be made. Some of the more common methods being used include decline curves, both rate-time and rate-cumulative, type curves based on analogy, numerical models, and advanced decline curve methods being proposed by the industry. What method does the SEC accept? The SEC should accept any method which gives reliable results. It should be demonstrated to be a "reliable technology" by the reserve auditor.

Reserve estimates should be basically the same for the SEC or PRMS guidelines since they are both principle based systems.

How do we show the method selected to estimate reserves meets the SEC's or PRMS's criteria? One way to demonstrate the methodology is reliable is to compare the EUR form year to year for a group of wells.

Some do not consider the Arp's Model to be appropriate for unconventional reservoirs. The problems include transient flow for much of the producing life of the well and a changing 'b'. High 'b' values – greater than one – are observed. Often a final minimum exponential decline is used for the last segment of the forecast to account for this. The minimum decline is normally based on analogy. If analogy shows a value for 'b' different than zero, than this can also be used.

Fetkovich has developed type curves which model the early transient data and then go to the Arp's model when boundary dominated flow is reached.

The Stretched Exponential Decline Model (SEDM) has also been suggested as a tool for reserve estimates in reservoirs which have both transient and boundary dominated flow regimes. It is thought to give more conservative estimates than ARP's and to be more accurate.

Some engineers prefer using a Linear Flow Model followed by an exponential decline when boundary dominated flow is reached. The final decline trend is based on analogy.

There is the Duong Model as well which also attempts to account for the long linear flow regime seen in unconventional wells.

The EUR for unconventional wells will depend on the model selected to estimate reserves. There is at present no consensus on which model is the best and should be used. Care should be used that the most appropriate model is used and the reason it is being used articulated.

Resource Plays

The Resource Play has been described as a continuous hydrocarbon deposit over a large area.[264] Resource plays can be oil or gas although most to date have been gas producing.

The Society of Petroleum Evaluation Engineers (SPEE) describes four criteria for Resource Plays:

1. A repeatable statistical distribution of estimated ultimate recoveries (EURs)
2. Offset wells are not a reliable indicator of the undeveloped location's performance
3. It has a continuous hydrocarbon system of regional extent
4. Free hydrocarbons are not held in place by hydrodynamics[265]

264 Society of Petroleum Evaluation Engineers Monograph 3; Guidelines for the Practical Evaluation of Undeveloped Reserves in Resource Plays; p.3.

265 Id.

The area required for a resource play and the development of a usable statistical model will most likely both require a minimum of 100 wells.[266]

The SPEE goes on to describe 4 additional criteria commonly seen in resource plays:[267]

1. Stimulation required to produce
2. Little water production
3. No obvious trap
4. Low permeability

These last four criteria are not required for the reservoir to be considered a Resource Play, but they are commonly seen in this type of reservoir.

Although similar, the SEC and SPE-PRMS definitions may result in different reserve numbers. The SEC reserve definitions state that "The term economically producible, as it relates to a resource, means a resource which generates revenue that exceeds, or is reasonably expected to exceed, the costs of the operation."[268] "Existing economic conditions include prices and costs at which economic producibility from a reservoir is to be determined. The price shall be the average price during the 12-month period prior to the ending date of the period covered by the report, determined as an unweighted arithmetic average of the first-day-of-the-month price for each month within such period, unless prices are defined by contractual arrangements, excluding escalations based upon future conditions."[269]

The SPE says "Resources evaluations are based on estimates of future production and the associated cash flow schedules for each development project...While each organization may define specific investment criteria, a

266 Society of Petroleum Evaluation Engineers Monograph 3; Guidelines for the Practical Evaluation of Undeveloped Reserves in Resource Plays; p.4.

267 Id.

268 Securities and Exchange Commission, Modernization of Oil and Gas Reporting, December 31, 2008; Sec. 210.4-10(a)(10).

269 Securities and Exchange Commission, Modernization of Oil and Gas Reporting, December 31, 2008; Sec. 210.4-10(a)(22)(v).

project is generally considered to be "economic" if its "best estimate" case has a positive net present value under the organization's standard discount rate, or if at least has a positive undiscounted cash flow."[270] "The economic evaluation underlying the investment decision is based on the entity's reasonable forecast of future conditions, including costs and prices, which will exist during the life of the project (forecast case). Such forecasts are based on projected changes to current conditions; SPE defines current conditions as the average of those existing during the previous 12 months. Alternative economic scenarios are considered in the decision process and, in some cases, to supplement reporting requirements. Evaluators may examine a case in which current conditions are held constant (no inflation or deflation) throughout the project life (constant case)."[271]

The SEC bases its economic parameters on the historical average and prices and costs are held constant going forward – unless adjusted by contract. The SPE is forward looking and permits changes in costs and prices over the life of the project. The SEC and SPE cases are equivalent if the costs and prices are held constant over the life of the project.

The SPEE presents four steps to estimate reserves for locations in a Resource Play:

1. Identify analogy wells
2. Create a statistical distribution for those wells
3. Determine the number of locations
4. Calculate the reserves using appropriate definitions

Analogous wells should be comparable in regards to the time period during which they were drilled, the technology used to drill and complete the wells, and the geology.[272] All three of these have an impact on the well and how it produces.

270 SPE-PRMS 3.1.1.

271 SPE-PRMS 3.1.2.

272 Society of Petroleum Evaluation Engineers Monograph 3; Guidelines for the Practical Evaluation of Undeveloped Reserves in Resource Plays; p.57.

Offset locations are often based on one location offsets. In that case, one proved producing well can have eight proved undeveloped locations surrounding it, 16 probable locations and 24 possible locations for a total of 48 locations offsetting the producing well. The SEC says "Reserves on undrilled acreage shall be limited to those directly offsetting development spacing areas that are reasonably certain of production when drilled, unless evidence using reliable technology exists that establishes reasonable certainty of economic producibility at greater distances."[273]

In reviewing undeveloped locations there are elements to consider in addition to economics and spacing. One must also consider the development plan and the timing of the drilling, project approval is required and the project funding must be available.

A development plan is required before assigning reserves to locations since the project must be defined enough to determine its commercial viability and to insure needed approvals and funding are in place along with the firm intent to proceed with development within a reasonable time frame. The SPE-PRMS says "To be included in the Reserves class, a project must be sufficiently defined to establish its commercial viability. There must be a reasonable expectation that all required internal and external approvals will be forthcoming, and there is evidence of firm intention to proceed with development within a reasonable time frame."[274]

The SEC takes a similar approach and requires a development plan and intent to develop but also includes five years as the time frame in which the locations must be drilled. "Undrilled locations can be classified as having undeveloped reserves only if a development plan has been adopted indicating that they are scheduled to be drilled within five years, unless the specific circumstances,

---

273 Securities and Exchange Commission, Modernization of Oil and Gas Reporting, December 31, 2008; Sec. 210.4-Definitions (31)(i)

274 SPE-PRMS Table 1

justify a longer time."[275] The PRMS suggests five years as a benchmark. "A reasonable time frame for the initiation of development depends on the specific circumstances and varies according to the scope of the project. While 5 years is recommended as a benchmark, a longer time frame could be applied where, for example, development of economic projects are deferred at the option of the producer for, among other things, market-related reasons, or to meet contractual or strategic objectives. In all cases, the justification for classification as Reserves should be clearly documented."[276] The SEC and PRMS do not always agree what a project is. The SEC may consider each location a project and so some wells may fall outside the five year rule and be disallowed although field development has begun and is proceeding.

The SEC has accepted horizontal locations parallel to horizontal producers, but has not taken a positive view of toe and heel locations.

The SEC leaves the choice of a method to estimate reserves to the auditor. Various methods are used by different evaluators and include analogy, DCA, advanced DCA methods, volumetric estimates and simulation. As well evaluations are updated, consistency of results should be considered. If the reserves are increasing or staying the same, the method used should be acceptable to the SEC to estimate proved reserves. Methods with show less consistency can be used for 2P and 3P reserve estimates.

Decline curve analysis in Resource Plays is difficult, especially early in the life of the well. The decline trend of a well is dictated by rock and fluid properties as well as operational and producing conditions. The Arp's Equation normally used is based on boundary dominated and not linear flow. Shale plays or other types of Resource Plays often have a long period of linear flow and so results from using Arp's are often in error. Most of these types of wells start with a steep initial decline and transition into a shallower exponential decline. The

---

275 Securities and Exchange Commission, Modernization of Oil and Gas Reporting, December 31, 2008; Definitions (31)(ii)

276 SPE-PRMS Table 1

initial periods of steep decline are dominated by flow from the fractures and later when the declines are smaller production is dominated by flow from the matrix. The resulting b factor is often greater than 1.[277] Many different approaches have been taken to accommodate this unique situation. In general, a good estimate from decline curve analysis is not available until after the well has produced for a longer period of time than would be expected from a conventional reservoir, usually two to three years.

The b factor in the Arp's equation relates to the curvature of the forecast, it relates to the rate at which the decline rate is changing. When b is equal to zero, the decline rate is constant and the curve exhibits an exponential decline. A recent study of horizontal shale gas wells showed b factors in the range of 0.6377 in the Fayetteville Shale to 1.5933 in the Barnett Shale.[278] By comparison, the Cotton Valley vertical tight gas wells averaged a b factor of 1.2778.[279] Unless based on analogy, care should be taken not to use ARP's without some modifications, such as a minimum final decline, when b is greater than one. The Dmin can be based on analogy.

---

277 SPE 135555; Shale Gas Production Decline Trend Comparison Over Time and Basins; , J. Baihly, R. Altman, R. Malpani and F. Luo; Sept, 2010

278 Study Asses Shale Decline Rates, J. Baihly, R. Altman, R. Malpani and F. Luo; The American Oil and Gas Reproter, May 2011

279 Id.

Improved Recovery

Additional reserves based on improved recovery can be classified as proved according to the SEC definitions, if there is a successful pilot or the project is ongoing and performance indicates the additional reserves are justified.[280] The company may use an ongoing project in the same formation in the immediate area as an analogy to assign proved reserves if the formation properties in the new project are as good as or better than in the analogy.[281]

The SEC makes a distinction between pressure maintenance projects where injection is started early in the life of the project and secondary recovery projects.[282] They have stated that in limited situations, such as some North Sea fields, the distinction can be made. The SEC will review the data presented by the company to support the contention improved recovery can actually be achieved.

The SPE/WPC definitions require a successful pilot or analogy to a successful project in an analogous reservoir with similar rock and fluid properties for the assignment of

---

280 Rule 4-10(a-2ii) Regulation S-X, Securities and Exchange Act of 1934 "Reserves which can be produced economically through application of improved recovery techniques (such as fluid injection) are included in the "proved" classification when successful testing by a pilot project, or the operation of an installed program in the reservoir, provides support for the engineering analysis on which the project or program was based."

281 SEC Division of Corporation Finance: Frequently Requested Accounting and Financial Reporting Interpretations and Guidance, March 31, 2001 3(c) "...If an improved recovery technique which has not been verified by routine commercial use in the area is to be applied, the hydrocarbon volumes estimated to be recoverable cannot be classified as proved reserves unless the technique has been demonstrated to be technically and economically successful by a pilot project or installed program in that specific rock volume. Such demonstration should validate the feasibility study leading to the project."

282 Topic 12 of Accounting Series Release No. 257 of the Staff Accounting Bulletins Question 2: In determining whether "proved undeveloped reserves" encompass acreage on which fluid injection (or other improved recovery technique) is contemplated, is it appropriate to distinguish between (i) fluid injection used for pressure maintenance during the early life of a field and (ii) fluid injection used to effect secondary recovery when a field is in the late stages of depletion? The definition in Rule 4-10(a)(4) does not make this distinction between pressure maintenance activity and fluid injection undertaken for purposes of secondary recovery.
Interpretive Response: The Office of Engineering believes that the distinction identified in the above question may be appropriate in a few limited circumstances, such as in the case of certain fields in the North Sea. The staff will review estimates of proved reserves attributable to fluid injection in the light of the strength of the evidence presented by the registrant in support of a contention that enhanced recovery will be achieved.

reserves based on improved recovery.[283] The SPE/WPC definitions also include the statement there must be a reasonable certainty the project will proceed.

Improved recovery is discussed by the SPE in Section 2.3.4 Improved Recovery.[284] The SPE describes improved recovery as additional reserves from a naturally occurring reservoir by supplementing the natural reservoir performance.[285] It can be through waterflood, tertiary recover or any other method used to supplement the reservoir performance.

There should be an expectancy the project will be economic and management plans to proceed with the project within a reasonable time, generally no longer than five years.[286] The SPE requires a pilot project or analogy to show the project will be commercial.[287] If the improved recovery method has not been established through routine applications, it should be included as reserves only after a successful pilot or installed project.[288]

---

283 SPE/WPC Reserve Definitions "Reserves which are to be produced through the application of established improved recovery methods are included in the proved classification when (1) successful testing by a pilot project or favorable response of an installed program in the same or an analogous reservoir with similar rock and fluid properties provides support for the analysis on which the project was based, and, (2) it is reasonably certain that the project will proceed. Reserves to be recovered by improved recovery methods that have yet to be established through commercially successful applications are included in the proved classification only (1) after a favorable production response from the subject reservoir from either (a) a representative pilot or (b) an installed program where the response provides support for the analysis on which the project is based and (2) it is reasonably certain the project will proceed."

284 Petroleum Resources Management System, 2007, Sec2.3.4.

285 Petroleum Resources Management System, 2007, Sec2.3.4 "Improved recovery is the additional petroleum obtained, beyond primary recovery, from naturally occurring reservoirs by supplementing the natural reservoir performance. It includes waterflooding, secondary or tertiary recovery processes, and any other means of supplementing natural reservoir recovery processes."

286 Petroleum Resources Management System, 2007, Sec2.3.4 "There should be an expectation that the project will be economic and that the entity has committed to implement the project in a reasonable time frame (generally within 5 years; further delays should be clearly justified)."

287 Petroleum Resources Management System, 2007, Sec2.3.4 "The judgment on commerciality is based on pilot testing within the subject reservoir or by comparison to a reservoir with analogous rock and fluid properties and where a similar established improved recovery project has been successfully applied.
Incremental recoveries through improved recovery methods that have yet to be established through routine, commercially successful applications are included as Reserves only after a favorable production response from the subject reservoir from either (a) a representative pilot or (b) an installed program, where the response provides support for the analysis on which the project is based."

288 Petroleum Resources Management System, 2007, Sec2.3.4 "The judgment on commerciality is based on pilot testing within the subject reservoir or by comparison to a reservoir with analogous rock and fluid properties and where a similar established improved recovery project has been successfully applied. Incremental recoveries through improved recovery methods that have yet to be established through routine, commercially successful applications are included as Reserves only after a favorable production response from the subject reservoir from either (a) a representative pilot or (b) an installed program, where the response provides support for the analysis on which the project is based."

The SPE classifies improved recovery reserves as producing only after the project is in operation.[289] The improved recovery volumes are considered developed when the necessary equipment has been installed or the cost to do so is minor.[290] If additional facilities or major expenses are required before the project can start, it should be classified as undeveloped.[291]

The SEC says an improved recovery project can be included as proved reserves if the basis is a successful pilot in the formation, a successful project in an analogous field, or other reliable technology which establishes with a reasonable certainty the engineering analysis on which the project is based.[292] The SEC also requires approval of the project by all necessary parties including government entities.

The SEC also says no reserves should be set up unless there has been an actual successful project in the reservoir or analogous reservoirs or other evidence using reliable technology establishing reasonable certainty.[293]

---

289 Petroleum Resources Management System, 2007, Table 2: Reserves Status Definitions and Guidelines p. 28 "Improved recovery reserves are considered producing only after the improved recovery project is in operation."

290 Petroleum Resources Management System, 2007, Table 2: Reserves Status Definitions and Guidelines p. 28 "Improved recovery reserves are considered 'developed' only after the necessary equipment has been installed, or when the costs to do so are relatively minor compared to the cost of a well. Developed Reserves may be further sub-classified as Producing or Non-Producing."

291 Petroleum Resources Management System, 2007, Table 2: Reserves Status Definitions and Guidelines p. 28 "Undeveloped Reserves are quantities expected to be recovered through future investments: ... (4) where a relatively large expenditure (e.g. when compared to the cost of drilling a new well) is required to (a) recomplete an existing well or (b) install production or transportation facilities for primary or improved recovery projects."

292 Securities and Exchange Commission, Modernization of Oil and Gas Reporting, December 31, 2008; 210.4-10 (22) (iv) "Reserves which can be produced economically through application of improved recovery techniques (including, but not limited to, fluid injection) are included in the proved classification when: (A) Successful testing by a pilot project in an area of the reservoir with properties no more favorable than in the reservoir as a whole, the operation of an installed program in the reservoir or an analogous reservoir, or other evidence using reliable technology establishes the reasonable certainty of the engineering analysis on which the project or program was based; and (B) The project has been approved for development by all necessary parties and entities, including governmental entities."

293 Securities and Exchange Commission, Modernization of Oil and Gas Reporting, December 31, 2008; 210.4-10 (31) (iii) "Under no circumstances shall estimates for undeveloped reserves be attributable to any acreage for which an application of fluid injection or other improved recovery technique is contemplated, unless such techniques have been proved effective by actual projects in the same reservoir or an analogous reservoir, as defined in paragraph (a)(2) of this section, or by other evidence using reliable technology establishing reasonable certainty."

Probable Reserves

Probable reserves were not recognized by the SEC for reporting purposes prior to the December 31, 2008 revisions, but have been by the SPE/WPC. Unproved reserves can be dependent on data which may be unavailable at report time. Unproved reserves may also result from technical, contractual, economic or regulatory issues.

Guidelines for assigning probable reserves are given by the Society of Petroleum Evaluation Engineers and include the following points.[294] Probable reserves should be in a formation known to be productive in the general area. They may be assigned to formations which appear to be productive based on log data, but do not have a test, core data or are not analogous to proven reserves in the area. Subsurface information must indicate the location is within the productive outline of the reservoir. Reservoir continuity should be expected.

Probable reserves lack the certainty of proved reserves. Using deterministic methods, they are reserves that are more likely than not to be recovered, but do not have a reasonable certainty of being recovered. They are volumes expected to be moved into the proved category as more data becomes available. The event to occur which will enable these reserves to be classified as proved should be identified. Under the probabilistic methodology, the proved plus probable reserves should have at least a 50% chance of being recovered.[295]

The reasons and assumptions for assigning probable reserves should be documented. As mentioned above, the event needed to move reserves from probable to proved should also be documented.

---

294 Reserves Definitions Committee Society of Petroleum Evaluation Engineers, "Guidelines for Application of Petroleum Reserves Definitions", p.16.

295 SPE/WPC Petroleum Resource Management System 2007, 2.2.2 and Table 3, "An incremental category of estimated recoverable volumes associated with defined technical uncertainty. Probable Reserves are additional Reserves that are less certain to be recovered than Proved Reserves. It is equally likely that actual remaining quantities recovered will be greater or less than the sum of the estimated Proved plus Probable Reserves (2P). In this context, when probabilistic methods are used, there should be at least a 50% probability that the actual quantities recovered will equal or exceed the 2P estimate."

The SEC uses a similar definition of probable reserves as the PRMS, in that they are "...additional reserves that are less certain to be recovered than proved reserves but which, in sum with proved reserves, are as likely as not to be recovered."[296] When deterministic methods are used, it is as likely as not that the remaining reserves recovered will equal or exceed the proved plus probable estimate.[297] If probabilistic methods are used, there must be at least a 50% probability the proved plus probable reserves recovered will equal or exceed the proved plus probable estimate.[298]

The SEC requires reporting only proved reserves, but now allows probable reserves to be reported if the company so desires.

Examples of Probable Reserve

Reserves lacking the certainty required by the proved reserve definition may be included in the reserve report as probable reserves for some exchanges outside of the United States.[299] Under the new SEC guidelines, probable reserves can be reported but are not required to be reported.

296 Rule 4-10(a)(18) [17 CFR 210.4-10(a)(18)].

297 Securities and Exchange Commission, Modernization of Oil and Gas Reporting, December 31, 2008; p.38.

298 Id.

299 SPE/WPC Petroleum Resources Management System 2007, Table 1, "The 2P and 3P estimates may be based on reasonable alternative technical and commercial interpretations within the reservoir and/or subject project that are clearly documented, including comparisons to results in successful similar projects. In conventional accumulations, Probable and/or Possible Reserves may be assigned where geoscience and engineering data identify directly adjacent portions of a reservoir within the same accumulation that may be separated from Proved areas by minor faulting or other geological discontinuities and have not been penetrated by a wellbore but are interpreted to be in communication with the known (Proved) reservoir. Probable or Possible Reserves may be assigned to areas that are structurally higher than the Proved area. Possible (and in some cases, Probable) Reserves may be assigned to areas that are structurally lower than the adjacent Proved or 2P area. Caution should be exercised in assigning Reserves to adjacent reservoirs isolated by major, potentially sealing, faults until this reservoir is penetrated and evaluated as commercially productive. Justification for assigning Reserves in such cases should be clearly documented. Reserves should not be assigned to areas that are clearly separated from a known accumulation by non-productive reservoir (i.e., absence of reservoir, structurally low reservoir, or negative test results); such areas may contain Prospective Resources. In conventional accumulations, where drilling has defined a highest known oil (HKO) elevation and there exists the potential for an associated gas cap, Proved oil Reserves should only be assigned in the structurally higher portions of the reservoir if there is reasonable certainty that such portions are initially above bubble point pressure based on documented engineering analyses. Reservoir portions that do not meet this certainty may be assigned as Probable and Possible oil and/or gas based on reservoir fluid properties and pressure gradient interpretations."

The SEC guidance says probable reserves can be assigned in an area where no proved reserves have been assigned.[300] They say however that it should be done only in exceptional cases such as projects awaiting governmental approval or improved recovery projects awaiting response. In no case should reserves be assigned without a penetration in the reservoir if it separated from the known accumulation by a major, sealing fault.[301] Volumes lacking commercial producibility are not reserves of any category.[302]

Probable reserves include volumetric estimates of volumes below the lowest known hydrocarbon. They are generally limited to a volume one sand thickness below the lowest known hydrocarbon. Under the SEC guidelines for probable reserves, the down dip volume cannot be assigned reserves unless there is data below the low known hydrocarbon suggesting hydrocarbons exist below the low known hydrocarbon.[303]

Probable reserves may also involve volumes not on leases owned by the company, but expected to be recovered by their well. Up dip volumes fall into the probable category if the drive mechanism is unknown.

Probable reserves also include volumes identified based on wireline logs, but have no test or analogy, as required for the proved classification. Improved recovery projects that have not been started or do not have an analogy to an economic, successful project in the same area and reservoir are considered probable.

---

300 Compliance and Disclosure Interpretations, October 26, 2009 Question 117.02, "...disclosure of unproved reserves without associated proved reserves should be done only in exceptional cases, such as for (1) development projects where engineering, geological, marketing, financing and technical tasks have been completed, but final regulatory approval is lacking or (2) improved recovery projects, at or near primary depletion, that await production response. Reserves should not be assigned without well penetration of the subject reservoir (rock volume) in the contiguous area that yields technical information sufficient to support the attributed reserve category. Volumes that are not economically producible are not reserves of any classification and should not be disclosed."

301 Compliance and Disclosure Interpretations, October 26, 2009 Question 117.04, "Un-penetrated, pressure-separated fault blocks should not be considered to contain reserves of any category until penetrated by a well."

302 Compliance and Disclosure Interpretations, October 26,2009 Question 117.02

303 Compliance and Disclosure Interpretations, October 26, 2009, Question 117.03 "Probable reserves may be assigned if reliable technology and data exist that, in the judgment of the evaluator, support characterizing those reserves as probable reserves. If no data exists below LKH, no unproved reserves can be assigned."

Undeveloped locations more than one offset from a commercial well should be considered probable if continuity of production cannot be shown with reasonable certainty based on performance or tests. Locations requiring new spacing rules should be carried as probable unless the governmental agency in charge of the area has a history of automatically granting new spacing. Likewise, reserves to be produced beyond the life of a PSC should be considered probable until an extension or a new PSC has been granted.

Probable should be used for reserves based on higher recovery factors from material balance estimates. When a volumetric estimate is not applicable and performance estimates are not unique, the more optimistic decline curve estimates, through either a flatter decline or, if no decline is present, a longer flat period, are assigned probable reserves.

Reserves in untested zones, not produced in the field, should be classified probable if the logs are not definitive and there is no analogy. Reserves for completions, recompletions, or locations not budgeted should be in the probable, rather than proved, category.

The SEC requires a penetration in the reservoir before reserves of any category can be assigned.[304]

Enhanced recovery reserves should be considered probable if there is no analogy or there is not a successful pilot project.

China

Probable reserves (C-D) are defined as reserves attributed to reservoirs in which a wildcat well has been drilled and tested at commercial rates, but either too few appraisal wells have been drilled to adequately determine the reserves, or the rates are poor.[305] An evaluation shows the reserves to be economic. The error

304 Federal Register /Vol. 74, No. 9 /Wednesday, January 14, 2009 /Rules and Regulations, p. 2167.

305 The National Standard of P.R.C., Petroleum Reserve Standard.

associated with these reserves should be no more than plus or minus 50%.[306]

Probable reserves involve reservoirs in which drilling has delineated most of the reservoir. The reservoir boundaries and thickness are almost known with certainty. The reservoir information is reliable enough to base additional appraisal wells on but more data is required for additional medium and long range planning.[307]

Probable reserves also include reservoirs having appraisal wells drilled, but the wells are not drilling out as expected. The reservoir may be lower quality than expected or the flow rates less than expected and so the appraisal drilling is cancelled. If the appraisal project is cancelled due to economic reasons the reserves are classified as probable.[308] In a large project, part of the reserves, less than 30% of the total proved plus probable volume, may be classified as probable in order to reduce the proved investment risk.

The new PRC standards state Probable Economic Initially Recoverable Reserves have a 50% probability the quantities actually recovered will equal or exceed the estimate and the development is economic.[309]

306 The National Standard of P.R.C., Petroleum Reserve Standard.

307 Id.

308 Id.

309 GAO Ruiqi, LU Minggang, ZHA Quanheng, XIAO Deming, HU Yundong, China Petroleum Resources/Reserves Classification.

Possible Reserves

Like probable reserves, possible reserves were not recognized by the SEC until the December 31, 2008 revision, but have been by the SPE/WPC.

The SEC definition says "... possible reserves include those additional reserves that are less certain to be recovered than probable reserves."[310] Possible reserves have little likelihood of actually being more than the estimated sum of proved, probable and possible reserves when deterministic methods are used.[311] Like the PRMS, the SEC says that for probabilistic estimates, there must be at least a 10% likelihood that the recovered reserves will equal or exceed the probabilistic estimate.[312] Like probable reserves, the SEC allows possible reserves to be reported, but does not require it.

Possible reserves are unproved reserves which geologic and engineering data indicate have less certainty than probable reserves. If probabilistic methods are used, there must be at least a 10% certainty of recovering at least that volume of reserves.[313] These reserves are not required to be reported.

The Society of Petroleum Evaluation Engineers gives guidelines for possible reserves.[314] Possible reserves should be associated known accumulations. They may appear productive based on log data but do not produce at commercial rates based on current prices and costs. They may be assigned if there is an indication of petroleum based on logs, mud logs, cores, or tests.

---

310 Securities and Exchange Commission, Modernization of Oil and Gas Reporting, December 31, 2008; p.39. Also see Rule 4-10(a)(17) [17 CFR 210.4-10(a)(17)].

311 Id.

312 Id.

313 SPE/WPC Petroleum Resources Management System 2007, 2.1.3.1 and Table 1, "An incremental category of estimated recoverable volumes associated with a defined degree of uncertainty. Possible Reserves are those additional reserves which analysis of geoscience and engineering data suggest are less likely to be recoverable than Probable Reserves. The total quantities ultimately recovered from the project have a low probability to exceed the sum of Proved plus Probable plus Possible (3P), which is equivalent to the high estimate scenario. When probabilistic methods are used, there should be at least a 10% probability that the actual quantities recovered will equal or exceed the 3P estimate."

314 Reserves Definitions Committee Society of Petroleum Evaluation Engineers, "Guidelines for Application of Petroleum Reserves Definitions", p.17.

Examples of Possible Reserves

The SPE gives examples of what they consider possible reserves.[315]

Possible reserves include volumes more than one sand thickness below the lowest known hydrocarbon. They may be based on seismic amplitudes or other available data. They also consist of volumes in zones which have been tested, but tests suggest they may not produce at commercial rates. If there are no tests, zones with too much uncertainty to be considered probable based on the wireline logs should be classified as possible.

Possible reserves are assigned to enhanced recovery projects without a successful pilot, if reservoir data or analogy suggests the project may not produce at commercial rates. Possible reserves are appropriate if facilities are not in place and are not budgeted.

Well locations beyond those set up as probable should be classified possible. Likewise, locations which are not legal locations based on current spacing rules, and the appropriate governmental agency has no history of automatically approving new spacing rules, are assigned possible reserves.

Areas dependent solely on seismic amplitudes, without geologic support, should be considered possible. Likewise, zones with questionable log characteristics and untested fault blocks that do not qualify as probable are classified as possible.

The SEC says

> "Possible reserves may be assigned where geoscience and engineering data identify directly adjacent portions of a reservoir within the same accumulation that may be separated from proved areas by faults with

315 SPE/WPC Reserve Definitions "In general, possible reserves may include (1) reserves which, based on geological interpretations, could possibly exist beyond areas classified as probable, (2) reserves in formations that appear to be petroleum bearing based on log and core analysis but may not be productive at commercial rates, (3) incremental reserves attributed to infill drilling that are subject to technical uncertainty, (4) reserves attributed to improved recovery methods when (a) a project or pilot is planned but not in operation and (b) rock, fluid, and reservoir characteristics are such that a reasonable doubt exists that the project will be commercial, and (5) reserves in an area of the formation that appears to be separated from the proved area by faulting and geological interpretation indicates the subject area is structurally lower than the proved area."

displacement less than formation thickness or other geological discontinuities and that have not been penetrated by a wellbore, and the registrant believes that such adjacent portions are in communication with the known (proved) reservoir."[316]

As with probable reserves, the SEC says possible reserves can be assigned below the LKH if the other criteria for reserves are met. They also require there be some data below the LKH before reserves are booked. If there is no data, then possible reserves cannot be booked below the LKH.[317]

The SEC says possible reserves can be booked without proved reserves in certain circumstances, but these are the exception rather than the rule.[318] They give the examples of development projects which lack governmental approval and improved recovery projects which have not had a response yet. Reserves should not be assigned without a well penetration in the reservoir or in the contiguous area. Volumes which are not economically producible should not be assigned as reserves.[319]

The SEC's new guidelines say no reserves of any category can be booked to a reservoir which does not have a well penetration if separated by pressure sealing faults from known accumulations.[320]

China

Possible reserves (D-E) are assigned to volumes on structures which are

316 Federal Register /Vol. 74, No. 9 / January 14, 2009 /Rules and Regulations) p. 2191, "Possible reserves may be assigned where geoscience and engineering data identify directly adjacent portions of a reservoir within the same accumulation that may be separated from proved areas by faults with displacement less than formation thickness or other geological discontinuities and that have not been penetrated by a wellbore, and the registrant believes that such adjacent portions are in communication with the known (proved) reservoir. Possible reserves may be assigned to areas that are structurally higher or lower than the proved area if these areas are in communication with the proved reservoir."

317 Compliance and Disclosure Interpretations, October 26, 2009, Question 117.01

318 Compliance and Disclosure Interpretations, October 26, 2009, Question 117.02

319 Id.

320 Compliance and Disclosure Interpretations, October 26, 2009, Question 117.04

delineated by seismic and have a wildcat well which tested at commercial rates. The variation in reservoir thickness is not known. Gas/oil contacts may not be known. Possible reserves form the basis of the evaluation project.[321]

Possible reserves can also include quantities, which although geologically similar to probable tested volumes, are not yet tested. Also, if the zone penetrated is not the major pay zone but has shows, the reserves are classified as possible.[322]

Technical operations may have been implemented and the future production should be greater than or equal to 10% of the estimated volume.[323]

321 The National Standard of P.R.C., Petroleum Reserve Standard.

322 Id.

323 GAO Ruiqi, LU Minggang, ZHA Quanheng, XIAO Deming, HU Yundong, China Petroleum Resources/ Reserves Classification.

SEC Reporting Requirements

General

The SEC disclosure requirements for oil and gas activities are codified in Subpart 1200 of Regulation S-K. Subpart 1200 includes much of Industry Guide 2 and additional disclosure requirements.[324]

With the December 31, 2008 revisions, the SEC has added certain reporting requirements.[325] They require the disclosure of non-traditional reserves such as bitumen, shale and coal as oil and gas reserves. The SEC now allows the reporting of probable and possible reserves. These are not required to be reported, but may be if the company desires. For instance, some large acreage holdings may have significant amounts of probable or possible reserves, but under the SEC guidelines have only limited amounts of proved oil and gas reserves. This is more in line with the PRMS reporting requirements. The SEC also now allows the disclosure of price sensitivity of reserves. Again, the company is not required to report this, but may if they chose.

The SEC requires the disclosure of proved undeveloped reserves, development plans and the technologies used to establish additional reserves. The company must report its internal controls over the reserve estimation process and the qualifications of the reserve estimator or auditor. The reserves must be reported by geographic area.

The SEC also added Item 1201 to Regulation S-K with the December 31, 2008 revision.[326] It states the company may modify the format of the tabular reporting specified in Item 1200 for ease of presentation, to add information, or to combine two or more required tables. Item 1201 also defines the term "geographic area".

---

324 Securities and Exchange Commission, Modernization of Oil and Gas Reporting, December 31, 2008; p.51.

325 Id.

326 Id.

Geographic Area

Item 1201 of Regulation S-K provides the definition of "geographic area". There was much discussion as to how much detail should be required to provide significant disclosure and yet provide the companies with some measure of security. The SEC concluded it should provide for meaningful disclosure under each company's unique situation by country, by groups of countries within a continent, or by continent.[327] The definition is similar to the one in SFAS 69. With the December 31, 2008 revisions, the SEC is providing percentage thresholds for the breakdowns.[328]

The December 31, 2008 guidelines provide for the disclosure of each country having 15% or more of the companies proved reserves, unless prohibited by that country.[329] The disclosure is based on the total worldwide proved oil and gas reserves of the company on a barrel of oil equivalent basis.[330]

Disclosure Rules

Under the December 31, 2008 guidelines, the SEC requires much of the disclosure information be presented in tabular form.

Disclosure of Reserves under Item 1202

A company's proved reserves were required to be reported under Instruction 3 to Item 102 of Regulation S-K. Item 1202 also requires disclosure of proved reserves and adds that probable and possible reserves can be disclosed.[331] It also permits disclosure of reserves from sources not considered traditional sources of oil and gas.[332] Item 1202 contains only

327 Securities and Exchange Commission, Modernization of Oil and Gas Reporting, December 31, 2008; p.55. and Item 1201(d) [17 CFR 229.1201(d)].

328 Securities and Exchange Commission, Modernization of Oil and Gas Reporting, December 31, 2008; p.55.

329 Id.

330 Item 1204(a) [17 CFR 229.1204(a)].

331 Item 1202 [17 CFR 229.1202]

332 Securities and Exchange Commission, Modernization of Oil and Gas Reporting, December 31, 2008; p.57.

one table for reporting oil and gas reserves with separate columns for each final product, such as oil, gas, synthetic oil, synthetic gas and other natural resources.[333]

Item 1202 requires disclosure both in aggregate and by geographic area for each product type and the following categories: proved developed, proved undeveloped, total proved, and optional reporting of probable developed, probable undeveloped, possible developed and possible undeveloped.[334] An example of the table format is in the SEC paper Modernization of Oil and Gas Reporting.[335]

If a company processes a natural resource such as bitumen into synthetic oil or gas prior to selling it, such reserves may be included under the synthetic oil and gas columns. The company can use the price of the synthetic oil or gas in determining whether or not it is economically producible.[336] If the natural resource is sold without processing, the reserves must be reported as other natural resources. Reserves cannot be based on the price of the final upgraded product.[337]

The reserves reported are the company totals of reserves determined for individual properties, wells, or reservoirs. Whether deterministic or probabilistic methods are used to estimate the reserves, the reported reserves are to be a simple arithmetic sum of the individual estimates within each reserve category.[338]

Item 1202 allows for the optional reporting of probable and possible reserves. If they are reported, they must be in the same format as the proved reserves and provide the same amount of detail as the geographic locations as required for proved reserves.[339]

The disclosure of estimates of oil and gas resources, and their value, which are not reserves, is prohibited. The exception to this is if these resources are

---

333 Securities and Exchange Commission, Modernization of Oil and Gas Reporting, December 31, 2008; p.57.

334 Securities and Exchange Commission, Modernization of Oil and Gas Reporting, December 31, 2008; p.58.

335 Id.

336 Securities and Exchange Commission, Modernization of Oil and Gas Reporting, December 31, 2008; p.59.

337 Id.

338 Securities and Exchange Commission, Modernization of Oil and Gas Reporting, December 31, 2008; p.60.

339 Id.

required to be reported under foreign or state law.[340] They can also be reported in the situation of a merger or acquisition if they were previously provided to the company wanting to merge or acquire the company.[341]

The rules provide reserves are to be reported using a 12 month average price. Item 1202 also provides for companies to disclose the sensitivity of their reserves to price fluctuations.[342] Different scenarios can be used, but the disclosure must also include the price and cost schedules and the assumptions used.[343] A sample of the format for this table is shown on page 66 of the SEC's Modernization of Oil and Gas Reporting.

The disclosure under Item 1202 is based on the end product sold.[344] Item 1202 requires the reporting of oil or gas separately from synthetic oil or gas. Under Item 1202, synthetic oil or gas require processing either while in the ground or after extraction before it can be used as oil or gas.[345] This would include bitumen and coal liquefaction or gasification, but not accumulations such as coalbed methane.

The SEC now requires the disclosure of certain qualifications of the primary technical person in charge of the reserve estimate and whether he is subject to a list if controls to maintain objectivity.[346] This disclosure does not apply to each person working on the report, but only to the person with primary responsibility for the report.

Although there was some discussion by the SEC to require companies to use outside third party reserve estimators, this requirement was deemed too onerous and use of third party consultants is not required. If a company does represent

---

340 Securities and Exchange Commission, Modernization of Oil and Gas Reporting, December 31, 2008; p.64.

341 Id.

342 Id.

343 Id.

344 Item 1202 [17 CFR 229.1202].

345 Securities and Exchange Commission, Modernization of Oil and Gas Reporting, December 31, 2008; p.67.

346 Securities and Exchange Commission, Modernization of Oil and Gas Reporting, December 31, 2008; p.68.

its reserves as being prepared by a third party, they must file a copy of the report as an exhibit to the registration statement.[347] The total report need not be filed, but only the summary of the work preformed and conclusions of the third party reserve estimator. The disclosure requirements are based on the SPE audit report guidelines and must include the purpose of the report and who it is prepared for, the effective date of the report, the percentage of the company's total reserves covered by the report and the geographical areas where these reserves are located and the assumptions, data, methods and procedures used in estimating the reserves. A discussion of the economic assumptions must be included. A discussion of the possible effect of regulations on the recovery of the stated reserves must be included. The risks and uncertainties of reserve estimates must be discussed. The preparer must state he has used all of the methods they consider appropriate and must sign the report.[348] Also if the report is for a reserve audit, it must contain the conclusions of the third party estimator. If it is related to a registration statement, it must have the

347 Securities and Exchange Commission, Modernization of Oil and Gas Reporting, December 31, 2008; p.71 and Item 1202(a)(8) [17 CFR 229.1202(a)(8)]. "Third party reports. If the registrant represents that a third party prepared, or conducted a reserves audit of, the registrant's reserves estimates, or any estimated valuation thereof, or conducted a process review, the registrant shall file a report of the third party as an exhibit to the relevant registration statement or other Commission filing."

348 Securities and Exchange Commission, Modernization of Oil and Gas Reporting, December 31, 2008; (8) Third party reports. If the registrant represent's that a third party prepared, or conducted a reserves audit of, the registrants reserves estimates, or any estimated valuation thereof, or conducted a process review, the registrant shall file a report of the third party as an exhibit to the relevant registration statement or other Commission filing. If the report relates to the preparation of, or a reserves audit of, the registrants reserves estimates, it must include the following disclosure, if applicable to the type of filing:

(i) The purpose for which the report was prepared and for whom it was prepared;

(ii) The effective date of the report and the date on which the report was completed;

(iii) The proportion of the registrant's total reserves covered by the report and the geographic area in which the covered reserves are located;

(iv) The assumptions, data, methods, and procedures used, including the percentage of the registrant's total reserves reviewed in connection with the preparation of the report, and a statement that such assumptions, data, methods, and procedures are appropriate for the purpose served by the report;

(v) A discussion of primary economic assumptions;

(vi) A discussion of the possible effects of regulation on the ability of the registrant to recover the estimated reserves;

(vii) A discussion regarding the inherent uncertainties of reserves estimates;

(viii) A statement that the third party has used all methods and procedures as it considered necessary under the circumstances to prepare the report;

(ix) A brief summary of the third party's conclusions with respect to the reserves estimates; and

(x) The signature of the third party.

(9) For purposes of this Item 1202, the term reserves audit means the process of reviewing certain of the pertinent facts interpreted and assumptions underlying a reserves estimate prepared by another party and the rendering of an opinion about the appropriateness of the methodologies employed, the adequacy and quality of the data relied upon, the depth and thoroughness of the reserves estimation process, the classification of reserves appropriate to the relevant definitions used, and the reasonableness of the estimated reserves quantities."

consent of the third party as an exhibit.

Item 1203 – Proved Undeveloped Reserves

The SEC requires disclosure of the "aging" or vintage of undeveloped reserves. Companies are now required to report proved undeveloped reserves converted to proved developed reserves during the year for the last five fiscal years. The reserves are to be reported by product type using current prices and costs. The SEC believes the company's track record is a better indicator than the forward looking statements showing the company plans for potential future development.

The SEC expressed concern that disclosure of certain information regarding undeveloped reserves may be misleading to investors if presented in a tabular form. They have therefore said in the December 31, 2008 guidelines the information must be presented in a narrative form and is to include the total amount of proved undeveloped reserves at year end, any material change in PUDs during the prior year including conversion into producing reserves, investments and movement during the year to convert the proved undeveloped reserves to proved developed reserves, and an explanation of why material amounts of proved undeveloped reserves have remained so for five years or longer.[349]

Item 1204 – Oil and Gas Production

In keeping with Industry Guide 2, Item 1204 requires reporting production by final product sold for the last three fiscal years. In addition, the company must disclose the average sales price and the average production cost per unit of production.[350] The disclosure must be by geographic area and on an oil equivalent barrels basis for each country and field containing 15% or more of the

349 Securities and Exchange Commission, Modernization of Oil and Gas Reporting, December 31, 2008; p.77-78.

350 Securities and Exchange Commission, Modernization of Oil and Gas Reporting, December 31, 2008; p.79.

company's proved reserves.[351]

If reserves are reported as "net before royalty" instead of "net production", this should be noted in the disclosure.[352]

Only the marketable production of natural gas should be reported. Flared, injected or gas used on lease are to be omitted. Gas-lift gas and reproduced gas should not be reported until sold.[353]

If other natural resources such as bitumen are sold before conversion to synthetic oil or gas, the production transfer prices and production cost should be reported separately.[354]

The transfer price and average production cost should be according to SFAS 69.[355]

Item 1205 – Drilling and Other Development Activities

There has been no change under the December 31, 2008 guidelines to Industry Guide 2 relating to drilling activities. The SEC has added a new rule requiring companies to discuss activities regarding resources extracted by mining techniques as these are now included under the definition of oil and gas producing activities.[356]

Item 1206 – Present Activities

There is no change to Item 7 of Industry Guide 2 under the December 31, 2008 revisions. The company's present activities must be disclosed, including the number of wells currently being drilled, waterfloods being installed, or any other

---

351 Securities and Exchange Commission, Modernization of Oil and Gas Reporting, December 31, 2008; p.80.

352 Id.

353 Id.

354 Id.

355 Securities and Exchange Commission, Modernization of Oil and Gas Reporting, December 31, 2008; p.81.

356 Securities and Exchange Commission, Modernization of Oil and Gas Reporting, December 31, 2008; p.82.

activities with material importance.[357]

Item 1207 – Delivery Commitments

The substance of the disclosures required by Item 8 of Industry Guide 2 is not changed with Item 1207. The company must report any arrangement calling for delivery of specified amounts of oil or gas and the steps the company is taking to meet those commitments.[358]

Materiality

Materiality is a term used in a number of places by the SEC. In general it means the information is important to the reader. The SEC does not place hard and fast numbers on what amount of change is material.

It defined in the U.S. GAAP as "…the magnitude of an omission or misstatement of accounting information that, in the light of surrounding circumstances, makes it probable that the judgment of a reasonable person relying on the information would have been changed or influenced by the omission or misstatement."[359]

The SEC says it is "…in substance identical to the formulation used by the courts in interpreting the federal securities laws. The Supreme Court has held that a fact is material if there is—a substantial likelihood that the...fact would have been viewed by the reasonable investor as having significantly altered the 'total mix' of information made available."[360]

The IASB defines materiality as well. It says "*Materiality*—Information is material if its omission or misstatement could influence the economic decisions

357 Securities and Exchange Commission, Modernization of Oil and Gas Reporting, December 31, 2008; p.83 and Item 1206 [17 CFR 229.1206].

358 Securities and Exchange Commission, Modernization of Oil and Gas Reporting, December 31, 2008; p.83 and Item 1207 [17 CFR 229.1207].

359 Speech by Roger Schwall at the SPEE Forum on SEC Reserves Definitions Oct. 22, 2002.

360 Id.

of users taken on the basis of the financial statements. Materiality depends on the size of the item or error judged in the particular circumstances of its omission or misstatement. Thus, materiality provides a threshold or cutoff point rather than being a primary qualitative characteristic which information must have if it is to be useful."[361]

The SEC says they are aware some companies have developed thresholds, such as 5%, for what they consider material. While they say they have no problem with this as a first look, all relevant data must be considered.[362] All matters required to give a fair and complete representation of the change should be considered.

What is material will vary in different situations. Items considered material in one instance may not be considered material in another. Items considered material one year may not be considered material the next year. In many circumstances, what is considered material is based on professional judgment.

361 Speech by Roger Schwall at the SPEE Forum on SEC Reserves Definitions Oct. 22, 2002.

362 Id.

SEC Reserve Reviews

A number of events can trigger SEC reserve reviews. The Sarbanes-Oxley Act requires a review of each public company at least once every three years. A reserve review can be initiated due to a request from a third party, or a whistle-blower reporting a problem with the company's reserve reporting. Reserve reviews can be the result of interest generated by press releases or other news items.

The SEC will review a company's 10-K for any unusual activity, such as major revisions of reserves or significant additions, either through discoveries or acquisitions. They will look at the relationship between proved developed and undeveloped reserves, and may question a disproportionate undeveloped reserve verses developed reserve base. They also have certain confidential screening criteria that are not discussed.[363] They have been targeting companies appearing to engage in "earnings management".[364]

The SEC will typically spend about five days reviewing a company.[365] The SEC will send a series of questions to the company being reviewed. The company will generally be given about three weeks to answer the questions. A series of letters may be involved. The SEC prefers not to meet face to face due to time constraints.[366]

The initial role of the SEC engineer was to review the classification of reserves, but in recent years the role has expanded to be much larger. The SEC engineer now has a significantly increased oversight role which includes screening for reservoir size and recovery factors, financial commitment, economic commerciality, and the availability of a market. The SEC has increased its

363 Speech by Roger Schwall at the SPEE Forum on SEC Reserves Definitions Oct. 22, 2002.

364 Id.

365 Id.

366 Id.

oversight into reserve reporting as can be seen from recent news articles.

On the technical side, the SEC engineers will look at how downdip limits are determined and whether or not they are below the lowest known hydrocarbon. They will look at how seismic has been used to define the reservoir. They will look at whether or not there was a flow test and if not, what data was used as the basis for proved reserves. They will review the data used to determine the reservoir parameters and to estimate the recovery factor. They will review undeveloped locations for legal locations and data supporting reservoir continuity. They inquire if there is a commitment to drill. If it is an enhanced recovery project, they will expect to find data supporting the projected increase in reserves, either the project is underway and showing a response or there is an analogy. They will also inquire if the facilities are in place or committed to. If reservoir simulation is used, they will review the data behind the assumptions and will examine the history match.

On the economic side, the SEC engineers will ask if year end pricing was used. They will review bookings under PSC's to see if these reserves are reported separately and if the life of the reserves extends past the life of the PSC. They will determine if there is a commitment to develop undeveloped reserves, and that the project has not stagnated. They will insure there is a market for non producing reserves. The SEC engineers will check for non-hydrocarbon revenue and for the proper handling of net profit interests.

Reserve Write-downs

If the SEC has a reserve review and feels reserves have been improperly classified as proved, they will request the company take a write-down and restate their reserves for the periods the improperly classified reserves were on the books. Again, Sarbanes-Oxley will have an impact if this is necessary, as the financial statements of the company have been certified by its officers as being correct. The Act provides penalties for misrepresenting the financials. Although engineers have always had a civil liability for the reserves they prepare, they will probably be held to a higher standard under Sarbanes-Oxley. What the additional liability will be has yet to be defined by the courts.

The recent reserve write-downs by various companies have centered on various issues noted above. There have been problems with undeveloped locations, such as not being legal locations or not having data to support the drainage areas assigned. There have been revisions as a result of using MDT data to extrapolate hydrocarbon downdip limits below the lowest known hydrocarbon. Seismic data used to extend the reservoir limit below the lowest known hydrocarbon, or updip of well control, or thicken the reservoir beyond what is seen in the wells, has resulted in reserve write-downs. Seismic cannot be used to assign proved reserves to untested fault blocks or analogous structures in the area of a producing structure. Reserve revisions have been the result of recovery factors being too high with not enough data on hand to support them. The SEC has taken the position that the most inefficient drive mechanism should be assumed until there is data supporting the higher recovery factor. For a reservoir model, the SEC requires a good history match and a geologic model that meets the SEC guidelines. The geologic model must honor the lowest known hydrocarbon and the porosity and water saturation distribution seen in the wells. The SEC feels a model will often represent the expected case rather than the more conservative proved case.

The SEC has also required write-downs because of economic issues. They require using the price for the last day of the year and not the average price received for the month. This is not necessarily the price the company received, but the price they would have received if they had sold the oil or gas into the spot market on that day. If the gas is subject to a contract, than the contract price must be used. The only escalations permitted are those required by contract. The SEC requires that reserves not extend past the life of the PSC. They require an approved development plan and established markets.

Reserve Estimators and Auditors

General

A reserve estimator is defined by the SPE as one who is responsible for estimating reserves and other reserve information.[367] A reserve auditor is one who is responsible for the conduct of an audit of reserve information estimated by others.[368]

Any reserve estimate depends on the quality of the available data and the qualifications of the reserve estimator. Qualified personal have the formal academic training and work experience to make professional judgments about reserve quantities and classifications. They are given adequate time to do the type of review requested. They are provided quality data upon which to base their judgments.

Reserve estimators can work for the company and prepare internal reports or they can be independent of the company; or outside, third party reserve estimators. The perception seems to be an outside reserve estimator will provide a more unbiased and reliable reserve estimate. This of course depends on the reserve estimator and the company requesting the estimate. One of the concerns, even with the outside reserve estimator, is the amount of influence a company has on the reserve process.

The company should provide whatever data is relevant to the reserve process. If there are differences in reserves, the reserve estimator and the company engineer

---

367 Auditing Standards for Reserves, SPE, June 2001 Sec 2.2. "*Reserve Estimator.* A Reserve Estimator is a person who is designated to be in responsible charge for estimating and evaluating reserves and other Reserve Information. A Reserve Estimator either may personally make the estimates and evaluations of Reserve Information or may supervise and approve the estimation and evaluation thereof by others."

368 Auditing Standards for Reserves, SPE, June 2001 Sec 2.2. "*Reserve Auditor.* A Reserve Auditor is a person who is designated to be in responsible charge for the conduct of an audit with respect to Reserve Information estimated by others. A Reserve Auditor either may personally conduct an audit of Reserve Information or may supervise and approve the conduct of an audit thereof by others."

should meet to insure that the reserve estimator has all of the data to properly estimate the reserves. Company management should not, however, try to force an estimate on the estimator, outside or internal, which is not supported by data. Reserves should be based on information and not pressure from management or the promise of a bonus based on results of the reserve estimate.

If outside reserve evaluators are making the estimate, they should be independent of the company. The company requesting the report should not be a major part of their business. They should not own an interest in the company whose reserves they are evaluating or in the properties being evaluated. Their fee should not be based on the outcome of the estimate.

The reserve evaluator, whether internal or external, should be trained in the procedures involved and in industry accepted best practices. They should be ethical and have integrity. They should understand the geologic and engineering principles involved. They should also be cognizant of the applicable reserve definitions to be applied.

SEC

The SEC does not define who can be a reserve estimator or auditor nor does it provide for licensing as such. The SEC does state the Securities Act of 1934 provides for civil liability for every expert who has been named as having prepared or certified any part of the registration statement and this would include the reserve report.[369]

There has been discussion in the past about the impartiality of third party evaluators. Some feel they can be coerced into higher numbers in order to keep the job for another year. This may be true of some consultants, but is not the

369 SEC Division of Corporation Finance: Frequently Requested Accounting and Financial Reporting Interpretations and Guidance, March 31, 2001. 3(m) "The SEC staff reminds professionals engaged in the practice of reserve estimating and evaluation that the Securities Act of 1933 subjects to potential civil liability every expert who, with his or her consent, has been named as having prepared or certified any part of the registration statement, or as having prepared or certified any report or valuation used in connection with the registration statement. These experts include accountants, attorneys, engineers or appraisers."

rule. In Canada, rules have been instituted making the oil and gas company responsible for the reserves they submit, even if they were done by a third party consultant. While pressure may exist to assign higher numbers, reserve evaluators should note their mission is to assign an objective reserve number based on the data available. The job of the evaluator is to insure that reserves assigned meet the requirements of the definitions they have been instructed to use. They should remember their ultimate responsibility is to the company shareholders to produce a report incorporating all of the data and making unbiased assumptions and interpretations based on that data.

With the advent of Sarbanes-Oxley, the use of consultants to participate in the reserve review process will be more frequent. They will be called on to help define the procedures and policies for reserve reporting and to review the internal reserve review process. They will be asked to review reserve classifications or do reviews of certain fields. More companies will want an outside reserve report or audit to insure compliance with the SEC guidelines by the company's internal reserve evaluators.

SPE

The SPE has published standards for reserve evaluators.[370] It is an attempt by the society to provide guidelines for standards to be adhered to by reserve evaluators, both geologists and engineers. In the first section, they discuss the purpose of estimating reserves, they discuss certain definitions, they discuss the qualifications of reserve auditors and reserve estimators, they apply standards for independence and objectivity, and they discuss the standards to be used for estimating reserves and for auditing reserves.

The SPE says a person is qualified to be a reserve estimator if they have the education, training, and experience to exercise professional judgment and to be

370 SPE Auditing Standards for Reserves "At its June 2001 meeting, the SPE Board of Directors endorsed recommendations of the Society's Oil and Gas Reserves Committee for revisions to the 1977 Reserves document. The changes make the SPE voluntary standard compatible with the 1997 joint SPE/World Petroleum Congress definitions for petroleum reserves (resources)."

responsible for estimating reserves and other reserve information.[371] A person is qualified to be a reserve auditor if they have the education, training, and experience to be responsible for the audit of reserve information estimated by others.[372]

The SPE says before reserve information can be relied on, the users must be comfortable it was prepared in an unbiased and objective manner.[373] The reserve evaluator should be independent with respect to the reserves being reviewed and he should have no investment in the company or the reserves. The fees received should not be contingent on the result of the results of the work performed. The reserve evaluators should be reportable to management and not in a reserve group or operations group so they can objectively report the reserves and have the freedom to report any irregularities. The evaluator should keep all information and data of the company confidential unless the company consents to the release of that information.

371 Auditing Standards for Reserves, SPE, June 2001 SEC 3.2 "A Reserve Estimator shall be considered professionally qualified in such capacity if he or she has sufficient educational background, professional training and professional experience to enable him or her to exercise prudent professional judgment and to be in responsible charge in connection with the estimating of reserves and other Reserve Information."

372 Auditing Standards for Reserves, SPE, June 2001 SEC 3.3 "A Reserve Auditor shall be considered professionally qualified in such capacity if he or she has sufficient educational background, professional training and professional experience to enable him or her to exercise prudent professional judgment while acting in responsible charge for the conduct of an audit of Reserve Information estimated by others."

373 Auditing Standards for Reserves, SPE, June 2001 SEC 4.1 "In order that users of Reserve Information may be assured that the Reserve Information was estimated or audited in an unbiased and objective manner, it is important that Reserve Estimators and Reserve Auditors maintain, respectively, the levels of independence and objectivity set forth in this Article IV. The determination of the independence and objectivity of Reserve Estimators and Reserve Auditors should be made on a case-by-case basis."

International Reserve Definitions

Different oil producing countries use their own reserve definitions, even though the SPE/WPC has made efforts to standardize the definitions worldwide.

Australia

In 1995, the Australian Petroleum Production and Exploration Association (APPEA) published mandatory guidelines for oil and gas reserves reported to the Australian Stock Exchange to be used together with the SPE/WPC definitions. The guidelines specify production should have a 90% chance of exceeding the proved reserves reported and a 50% chance of exceeding the proved plus probable reserves reported. Possible reserves are not discussed. They provide guidance as to fuel use and flared gas saying reserves should be net of fuel and flare losses and after removal of inert substances.[374]

In Western Australia, reserves with insufficient information or lacking a development plan to be classified proved or probable should be classified as possible. The probability of recover for these reserves ranges from 10 to 50%.[375]

Former Soviet Union (FSU)

Unlike other reserve classification systems, the FSU system focuses on the maturity of development rather than the risk and economics of producing the reported reserves. There is also discussion referring to the FSU meaning the volume of oil and gas in place when reserves are referred to and recoverable reserves to mean those reserves recoverable by economic means.[376]

---

374 Cronquist, Chapman, Estimation and Classification of Reserves of Crude oil, Natural Gas, and Condensate, 2001.

375 Id.

376 Id.

Reserves in the FSU are classified as follows:[377]

Class A: Producing reserves which have fully described properties and drive mechanism and have no more than 10% uncertainty.

Class B: Non-producing reserves which have flow tests or cores showing they will produce economically and also have a development plan. These reserves should have no more than a 15% uncertainty.

Class C1: These reserves are the basis of step-out drilling programs. They are adjacent to reserves in the A or B Class or in areas having some wells shown to be commercially productive and others having positive log and core data. These reserves should have no more than 50% uncertainty.

Class C2: This class includes reserves in untested zones or fault blocks adjacent to reservoirs having a higher classification.

Class C3: This class is prospective resources in untested zones or in identified prospects in a known oil and gas producing area.

Class D1: These are projected resources in regions having commercial oil and gas production.

Class D2: These are anticipated resources in regions without commercial oil and gas production.

Norway

The Norwegian Petroleum Directorate classifications are based on the maturity of reserve development rather than the risk associated with recovering a stated reserve volume.

Class 1: producing reserves.

Class 2: reserves with an approved development plan.

Class 3: reserves in the late planning phase – the plan of development and operation has been approved or will be within two years.

Class 4: the plan of development and operation will be approved within 10 years.

377 Cronquist, Chapman, Estimation and Classification of Reserves of Crude oil, Natural Gas, and Condensate, 2001.

Class 5: reserves to be developed in the long term.

Class 6: development is not likely, or reserves designated as a small technical discovery.

Class 7: new discoveries with incomplete evaluation.

Class 8: resources based on possible future increases in the recovery factor.

Class 9: resources in prospects.

Class 10: resources in leads.

Class 11: unmapped resources – based on statistical analysis.

For each classification "low", "base" and "high" estimates are made. For the low estimate, the probability of recovery is P80 or P90. There is no probability of recovery for the base estimate. For the high estimate, the probability of recovery should equal or exceed P10 or P20.[378]

The state oil company of Norway, Statoil, classifies reserves as discovered and undiscovered. Reserves are resources which are discovered and can be recovered economically.

Statoil classifies reserves as follows:[379]

Low estimate: 90% certainty the reserve estimate will be met or exceeded.

Expectation: expected final reserve recovery.

High estimate: There is a 10% certainty the reserve estimate will be met or exceeded.

United Kingdom

The United Kingdom requires commercial reserves to be reported as either a) proven and probable reserves or b) proved developed and undeveloped reserves.[380] The definitions and wording used to explain proved reserves are

378 Cronquist, Chapman, Estimation and Classification of Reserves of Crude oil, Natural Gas, and Condensate, 2001.

379 Id.

380 Statement of Recommended Practice, Accounting for Oil and Gas Exploration, Development, Production and Decommissioning. Updated 7th June 2001.

very similar to the SEC language. The SORP pertaining to reserve definitions is included as Appendix L.

The UK says reserves reported under a) above, are estimated quantities of oil or gas that geologic and engineering data demonstrates with a specified degree of certainty (50% probability the recovery will be greater than the proved plus probable estimate) to be recovered in future years from known reservoirs and are commercially producible.[381]

Reserves reported as proved developed and undeveloped reserves are estimated quantities of oil or gas which engineering and geological data indicate will be recovered in future years from known reservoirs under existing economic and operating conditions using the costs and prices at the date the estimate is made.[382]

Canada

The requirements for Canadian reserve reporting are spelled out in National Instrument (NI) 51-101. Although these requirements are similar to the SEC

381 "Proven and probable reserves are the estimated quantities of crude oil, natural gas and natural gas liquids which geological, geophysical and engineering data demonstrate with a specified degree of certainty (see below) to be recoverable in future years from known reservoirs and which are considered commercially producible. There should be a 50 percent statistical probability that the actual quantity of recoverable reserves will be more than the amount estimated as proven and probable and a 50 percent statistical probability that it will be less. The equivalent statistical probabilities for the proven component of proven and probable reserves are 90 percent and 10 percent respectively.
Such reserves may be considered commercially producible if management has the intention of developing and producing them and such intention is based upon:
- a reasonable assessment of the future economics of such production;
- a reasonable expectation that there is a market for all or substantially all the expected hydrocarbon production; and evidence that the necessary production, transmission and transportation facilities are available or can be made available..."
".... Reserves may only be considered proven and probable if producibility is supported by either actual production or conclusive formation test. The area of reservoir considered proven includes (a) that portion delineated by drilling and defined by gas-oil and/or oil-water contacts, if any, or both, and (b) the immediately adjoining portions not yet drilled, but which can be reasonably judged as economically productive on the basis of available geophysical, geological and engineering data. In the absence of information on fluid contacts, the lowest known structural occurrence of hydrocarbons controls the lower proved limit of the reservoir." Statement of Recommended Practice, Accounting for Oil and Gas Exploration, Development, Production and Decommissioning. Updated 7th June 2001.

382 "(b) Proved developed and undeveloped oil and gas reserves
the estimated quantities of crude oil, natural gas and natural gas liquids which geological and engineering data demonstrate with reasonable certainty to be recoverable in future years from known reservoirs under existing economic and operating conditions, that is, prices and costs as at the date the estimate is made." Statement of Recommended Practice, Accounting for Oil and Gas Exploration, Development, Production and Decommissioning. Updated 7th June 2001.

requirements, they also differ in some respects.

Canada requires public companies to report proved and probable reserves on an annual basis. They may also report possible reserves, but are not required to do so.

Canada requires producers, except large producers, to have the reserve report certified by a third party.

Canada allows the use of deterministic numbers for reserve estimates and suggests a P90 certainty for proved reserves. Unlike the SEC, they allow the use of probable reserves in their reporting. They allow the reporting of reserves without proof of the ability to finance the planned development. They use the P90, P50 and P10 in a deterministic sense, unlike the SEC. It is in effect mixing probabilistic and deterministic methods.

Another difference between the Canadian and SEC requirements is Canada allows aggregation of proved reserves, while the SEC does not. As mentioned before, this may result in a higher number than would be allowed under SEC guidelines. The more properties included, the greater the effect. The SEC, on the other hand, requires a straightforward reconciliation for reserve reporting. The aggregation effect can result in the small producer and the large producer having different reserves for the same field and thus become a cause of potential confusion for the investors.

The Canadians allow booking improved recovery reserves based on analogy, as does the SEC. Canada limits proved reserves to direct offsets with good geologic control, much as the SEC. The Canadians are also limited to the low known hydrocarbons for booking proved reserves.

Mexico

Mexico calls for an internal reserve audit twice each year as well as an annual external certification. The internal audits are for planning purposes and the external review is for transparency and to satisfy the requirements of the internal financial markets.

Mexico uses the SEC reserve definitions for reporting proved reserves and the SPE/WPC definitions to report probable and possible reserves.

Middle East

The larger national oil companies have an internal annual reserve audit, but no external audit of their reserves. They use the SPE/WPC guidelines for proved, probable and possible reserves.

Trinidad and Tobago

Trinidad and Tobago use the SPE/WPC reserve definitions for proved, probable and possible reserves. They have an external gas reserve certification annually.

Venezuela

Venezuela reserve definitions include proved, probable, and possible primary and secondary reserves. Proved primary reserves include fields with commercial production or reservoirs with a reasonable certainty of commercial production.

Reserve Reports vs. Audits

The reserve report and reserve audit are the two primary review processes companies use to verify their internal reserve estimates. The reserve report is an independent study of the properties by a third party, while the reserve audit is the third party review of a reserve report done by others. An opinion is provided on the reported reserves. Another type of audit is a third party review of a company's processes and procedures used to estimate their reserves. The process review does not typically involve a review of reserve quantities or the calculations or raw data used in the estimate. This type of superficial review is generally not sufficient for investors or the various reporting agencies.

Reserve Reports

The reserve report is an independent review of the company's reserves. The reserve evaluator will begin with basic data and make an independent estimate of reserves and future revenue. Seismic and petrophysical data may be reviewed to develop volumetric estimates if there is not enough performance to make a reserve estimate from production and pressure data. He will forecast his reserve estimate and apply economics. He will determine the appropriate reserve classification based on the reporting standards of the applicable agencies and the reserve definitions being used. The reserve definitions being applied must be noted in the reserve report discussion letter.

The reserve report generally includes all of the properties of the company, but may be limited at the company's request to a smaller number. In some cases where there is a large number of small, low value or low net interest properties, the company may request that only the top 80% – 90% of the value of the reserves be reviewed.

The reserve estimator will publish a report containing his reserve estimate and forecast of future production rates and economics. Included along with the cash flow will be a report which explains the assumptions made, the reserve definitions used, the age of the data available, the source of the data, the purpose of the report, and any unusual terms that require an explanation to be clear.

The reserve report will typically contain a discussion letter, one-line summary tables and detail cash flow pages. The discussion letter will include a statement of the scope of the report, definitions used in the reserve estimates, what methods were used to estimate the reserves, the source of the data used for the report, the pressure base used, and the volume measurements used to report the oil and gas. The cash flow itself will normally include gross volumes and net sales volumes, prices, operating costs, development and abandonment costs, production taxes, future net income and discounted future net income.

An "SEC Reserve Report" is used by companies for their 10-K filings. It is intended to be a standardized estimate at year-end, limited to proved reserves with economics dictated by a one day price – the price on the last day of the year.

Audit Reports

A reserve audit is a review of the reasonableness of the reserves reported by a company in light of the procedures used, and data available. It is not an independent review by the reserve auditor, but a review of work done by someone else. In 1977, the SPE set out standards for reserve audits. These standards were updated in 2001. In general, the audit can cover the procedures used by the reserve evaluator, qualifications of the internal reserve estimators, the historical reserve revision trends, the ranking by size of the properties, the percentage of properties estimated by each method, performance or volumetric, whether volumetric or performance estimates were made, and significant changes from the last report.[383] It can vary in

383 Auditing Standards for Reserves, SPE, June 2001 Sec 6.4 (e) "An audit of the Reserve Information pertaining to an Entity should include a review of (i) the policies, procedures, documentation and guidelines of such Entity with respect to the estimation, review and approval of its Reserve Information; (ii) the qualifications of Reserve Estimators internally employed by such Entity; (iii) ratios of such Entity's reserves to annual production for, respectively, oil, gas and natural gas liquids; (iv) historical reserve and revision trends with respect to the oil and gas properties and interests of such Entity; (v) the ranking by size of properties or groups of properties with respect to estimates of reserves or the future net revenue from such reserves; (vi) the percentages of reserves estimated by each of the various methods set forth in Section 5.3 for estimating reserves; and (vii) the significant changes occurring in such Entity's reserves, other than from production, during the year with respect to which the audit is being prepared."

detail from an in-depth audit to a broad brush review of procedures. It reports on the reasonableness of the underlying reserves and whether or not they generally conform to accepted geological and engineering standards. It can include a review of economics and forecasts or just reserves. Although it usually entails only 80% to 90% of the company's reserves, it can include a review of everything.

In the December 31, 2008 guidelines, the SEC defines the term reserves audit as "the process of reviewing certain of the pertinent facts interpreted and assumptions made that have resulted in an estimate of reserves prepared by others and the rendering of an opinion about the appropriateness of the methodologies employed, the adequacy and quality of the data relied upon, the depth and thoroughness of the reserves estimation process, the classification of reserves appropriate to the relevant definitions used, and the reasonableness of the estimated reserves quantities."[384]

The general rule is that 80% of the company's value is in 20% of their properties. The scope of the audit can be increased to include all of the company's reserves as well as the costs, prices, future production forecasts and economics.

Once the reserve process is completed, the reserve auditor will compare his reserves to those of the company. If the total is within an agreed upon tolerance—generally about 10%—he will issue the audit letter. If the reserves are not within tolerance, he will request the company adjust their reserves before the letter is written. After the reserves agree within tolerance, the letter is issued. The audit letter will usually include a statement as to the quality of the data and the procedures used to estimate reserves and a comment on the reasonableness of the reserve estimate.

The auditor will not sign off on the reserve numbers, but only the reasonableness of the reserves, classifications and procedures used. The review will include an opinion on the data and the geological and engineering procedures used in the report. If it does not include a review of the economics, the letter will be limited to an opinion of the reserves only. The letter will express an opinion about the

384 Securities and Exchange Commission, Modernization of Oil and Gas Reporting, December 31, 2008; p.73.

reasonableness of the reserves as a whole and whether or not they conform to accepted geological and engineering practices. Examples of Audit Letters have been prepared by the SPE and are included as part of Appendix D. Since no economic indicators are presented in this type of report, it is generally not suitable for loans, fair market value, or litigation.

The audit does have its uses however, and in certain situations it is preferable to a reserve report as it can be much less expensive.

The audit procedure begins with the company's reserve report and a review by the auditor of the methods and procedures used. The audit company will assign personnel to review the reserves based on their experience in certain areas, as experience and analogy are important factors in deciding the reasonableness of the assigned reserves. If there are any questions or problems, the auditor will meet with the company personnel and verify the data and procedures used. Selected data will be checked depending on its relevance and the relative reserve rank of the property. The auditor will then decide if he can agree with the company's estimate as reasonable. If not, the reserve auditor will make his own reserve estimate.

An oil and gas company may wish to have their reserve process and procedures reviewed. The purpose of this type of review is to insure their reserves have been estimated in compliance with appropriate reserve definitions, personnel making the estimates are qualified and had an adequate amount of data available for the review. It may entail the presence of a third party consultant at the company internal reserve reviews. The auditor may review of the company's engineer's notes and methodology he used to estimate the reserves. The process review may also look into the qualifications of the company's reserve evaluation team and the data available and used by them in preparing the reserves.

In its December 31, 2008 guidelines, the SEC says that a process review can be helpful, but if a company discloses a third party has prepared one, it must disclose the details about the review.[385]

385 Securities and Exchange Commission, Modernization of Oil and Gas Reporting, December 31, 2008; p.75.

Reserve Risk Assessment

Reserve Risk

Because of the uncertainty involved in reserve estimates and the assumptions required, there is a certain amount of risk the reserves will be what was estimated and portrayed in the report. The SEC has attempted to mitigate that risk by requiring a high degree of certainty for reserves reported for financial purposes. There are several things, however, that any user of a reserve report should consider to help them understand the risk and make adjustments for it.

The user of a reserve report should note the date of the report, both the "as of date" and the date that was actually prepared. If either of these is old, the user should be extra careful in relying on the report. There may have been significant changes since the work was done. Knowing the producing characteristics of the areas where the reserves are located is important in analyzing how dated a report can be and still be useful, and how much value it has. It is also important to note the date the estimator had received his last data. If the data predates the report "as of date" by too much, there is risk of significant changes the reserve estimator is unaware of. Sometimes a report is "mechanically" updated so that it has a new "as of date", but no additional work has been done and no new data reviewed. The user should also consider whether it is a reserve report or an audit report.

The user should be aware of who did the report, whether it was done internally or by a third party. The qualifications of the reserve estimator should be inquired about, his familiarity with the appropriate definitions and areas where the reserves are located, as well as his experience level.

The report user should be familiar with the properties in the report, their maturity, and what percentages of the properties are developed and undeveloped. The accuracy of the reserve estimate is directly proportional to the maturity of the property. As more data is available, the spread between the high estimate and the low estimate of reserves becomes smaller. The divergence decreases as the reserve category moves from undeveloped, to non producing, to producing. The longer the wells have been on production, the lower the risk the estimate will be off significantly.

One should ask how the reserves were determined and what percentages of the reserves were by performance estimates and what percentage by volumetric estimate, analogy, or other methods. The reserves estimates by performance methods generally have less risk associated with them than those by volumetric estimates or other methods. One should also consider the concentration of value of the properties in the report, whether it is spread over a number of properties or concentrated in just a few. The concentration of value in a few properties or even in just one property a higher risk than having the value spread among a number of properties. If the value is too concentrated, the loss of one or two wells can have a significant impact, while if the value is spread out, the loss of one or two wells will not be devastating.

The user of the report should inquire if there are any significant changes that have taken place since the report was published or after the "as of date" of the report. For instance, data acquired after the end of the year cannot be used in SEC reports. Although both the company and the reserve estimator know of a major adjustment that should be made, the adjustment cannot be made in a report using SEC definitions since it is based on data acquired after the date of the report. These types of changes, though, can and should be discussed in the text of the report. The database available to the reserve estimator should be reviewed to determine the sufficiency and age of the data. All of the data related to the properties should be available to the reserve estimator and it should be current. Price and cost data should also be reviewed for reasonableness.

The areas where the reserves are located are important as they may have operational, cost, environmental, or political risk. If the reserves are located overseas, then the life of the PSC should be reviewed to insure the reserves do not extend past it. The availability of a market should be insured if the wells are not yet on production, as well as timing for first production.

The report user should note the proportion of undeveloped verses producing reserves. The producing reserves are generally the lowest risk reserve category, while undeveloped reserves are inherently more uncertain. The lower the undeveloped reserves in relation to the producing reserves, the less uncertainty there is in the report.

The report user should compare the reports over the last three years and note any significant revisions. Frequent significant revisions indicate a higher uncertainty in the reserves reported.

Economic Risk

Economic risk involves pricing risk, cost risk and political risk. All of these factors need to be evaluated and understood.

Pricing Risk

Hydrocarbon prices vary depending on the perceived supply and future demand for the product. They can vary significantly from year to year, depending on weather, usage or political uncertainties and can have a profound impact on the value of the oil and gas portfolio. The prices used should be analyzed in light of the reserve definitions used in the report. With SEC reporting, the price on the last day of the year is required. The SPE, however, allows the use of an average price received.

Cost Risk

The cost of a well or project can often exceed the original estimate. This results in a project that may well be uneconomic or provide an unacceptable return to the investors. Before the project is started, a reasonable cost estimate should be prepared by competent people having access to current data. If the original estimate is no good, the project economics will not be valid.

Cost risk also involves the amount of time it takes to actually complete the project. If the work takes longer than expected due to poor planning or other factors unforeseen at the beginning of the project, the cost will be higher than anticipated. Often there are unforeseen events that prolong a project so a miscellaneous contingency amount should be included in the estimate. Many companies add 15% to the total cost of the work to take this into account. The cost estimated and the amount of time projected to complete the project should be compared to actual recent projects in the area for reasonableness. Assurances that this one will be cheaper or completed faster, should be analyzed and the reasons that this might happen should be documented and explained.

There is also the risk that production will not start when expected, either due to the facility and pipeline installation taking longer than planned, or there is no market available. This reduces current cash flow and the total present value of the project.

Political Risk

In many countries the minerals belong to the state and not the landowner. Although this makes it easier in some ways, it adds another element of risk to the project. In countries with unstable governments, the political climate can change with each election or coup. In the past, regime changes have resulted in nationalization of a company's assets and production. Changes in government leadership can also lead to different contract terms or a change in working relationships. Market is also a part of this risk especially if the government controls the market or access to it.

SPEE

The SPEE has assigned ranges for monetary value risk factors (MVRFs).[386]

386 Reserves Definitions Committee Society of Petroleum Evaluation Engineers, "Guidelines for Application of Petroleum Reserves Definitions", p.17.

They are considered to include economic considerations that will influence the resulting cash flow. They are given as follows:

- Proved Producing 0.85 to 1.0
- Proved Shut-In 0.65 to 1.0
- Proved Behind Pipe 0.5 to 0.9
- Proved Undeveloped 0.25 to 0.75
- Probable Behind Pipe 0.05 to 0.5
- Probable Undeveloped 0.00 to 0.4
- Possible Behind Pipe 0.00 to 0.25
- Possible Undeveloped 0.00 to 0.2

Sarbanes-Oxley

The Sarbanes-Oxley Act (Act), enacted in 2002, is a comprehensive revision of federal securities law. It amends the Securities Exchange Acts of 1933 and 1934, and also sets out additional provisions:

It establishes the Public Company Accounting Oversight Board to regulate and oversee accounting firms that audit public companies,
it amends the 1934 act and prohibits accounting firms from performing certain non-audit services at the same time as an audit,
it requires an audit committee,
it requires the executive and financial officers to certify financial reports,
it restricts trading by insiders, and
it increases the penalties for violations of the federal securities laws.

The six most important categories of provisions concern:

the Public Accounting Oversight Board,
auditor independence,
corporate responsibility,
financial disclosures,
conflicts of interest, and
penalties.

The boards of directors and the audit committees have been tasked with assuring the stockholders and other interested parties of the integrity of the information reported to them. The information needs to provide a clear understanding of the companies' financials.

The Act is an attempt to provide guidelines for minimum standards of integrity and ethical conduct of those directing public corporations, and penalties for those who do not follow the guidelines. The SEC has also enacted guidelines for corporate ethical conduct and a corporate code of ethics. However, laws or rules will not insure that individuals will act in an ethical manner. The best that can be done is to establish guidelines for what the expected behavior should be and report unethical behavior, behavior that does not meet those guidelines. The Act applies to public companies, domestic or foreign, although the SEC has the authority to exempt foreign corporations.

All three stock exchanges now have rules concerning outside directors on company boards. The New York Stock Exchange has passed rules requiring the majority of the board to come from outside the company. The boards also are now required to have regular sessions that insiders are not permitted to attend. Having board meetings without the influence of management or the CEO enables the board to act in a more independent manner, to speak more freely, and raise controversial issues.

As is the case with most new legislation, there are still many unanswered questions about the impact the Act will have on public, and by association, private corporations. Although much has been written about the Act, about its stated purpose of providing shareholders of public companies more reliable information to base their business decisions on, and about insuring more integrity and accountability for the board and the executive officers, the detailed workings of the Act may not be known until it has been in operation for some period of time, final rules have been implemented, and tested in the courts.

Effect on Private Companies

The board of directors of any company, public or private, owes a duty to the shareholders to advise management on important corporate decisions and to insure that management is performing their duties in a manner that protects the interests of the shareholders. The board of directors sets the tone for the company.

Best Practices

Sarbanes-Oxley as written is intended to address and reform the accounting practices involving public companies. Some feel, however, that even private companies will ultimately feel the effects of the Act as the principles put forth by it will eventually spread and be considered "best practices" by the auditors even when the company is not public and technically not under the rules of the Act. Also, lenders and investors may insist that certain rules of the Act be adapted in order for them to have a relationship with the private company. They may require an independent audit committee and directors, and require that the CEO and CFO certify the financial statements. Ethics programs may be desired and requested.

Board of Directors and Officers

The duties of care and loyalty are the same for the directors of a private company as they are for a public company. Since the passage of the Act, the directors of the public companies are being held to a higher standard than they once were especially in regard to auditor independence, financial statement certification and insider trading. It is probable that directors and officers of private companies will also be held to these higher standards. The potential liabilities of directors and officers of private companies likely will increase. The shareholders of the private companies will also probably expect more transparency in financial statements and management compensation matters.

Restrictions on Corporate Activities

Insider Trades

Section 306 makes it unlawful for any director or executive officer to trade that companies' securities during a pension blackout period.

Conflict of Interest

Section 402 makes it unlawful for a public corporation to make a personal loan to a director or executive officer of the corporation.

Equity Compensation Plans

The SEC has approved the NASDAQ and NSE rules relating to equity compensation. The companies listed on these exchanges now must have stockholder approval for equity compensation plans.[387] These plans are defined as any arrangement that calls for equity securities being transferred to an employee, director, or service provider as compensation including the granting of options even if not under a plan. Stock options granted upon initial employment are exempt from the rules, as are some retirement type funds.

The AMEX does not currently have the same rules restricting equity compensation without shareholder approval, but they have filed a proposed rule change which addresses the issue of stockholder approval for stock option and equity compensation plans.[388]

Corporate Transparency

Public Company Accounting Oversight Board

Purpose

Section 101 of the Act establishes the Public Company Accounting Oversight Board (Board). The Board is to watch over the auditors of public companies. All accounting firms, both U.S. and foreign, which prepare audits for public companies, are required to register with the Board by Section 102. Section 103, directs the Board to establish standards for the accounting firms and rules for retention of work papers and partner review. It also establishes rules for testing by the auditor of the company's internal controls as well as quality control issues relating to the audit itself and the auditing firm.

---

387 Release No. 34-48108; File Nos. SR-NYSE-2002-46 and SR-NASD-2002-140

388 File No. SR-Amex-2003-42.

Investigations

Section 105 of the Act gives the Board authority to conduct investigations. The Board can require the accountants or "any other person" to testify and produce audit work papers. If the Board determines that the accounting firm or an associated person has violated the Act, Board rules, SEC rules, or professional standards, it can impose sanctions. These sanctions range from requiring additional training to fines.

Review of Company Reports

Section 104 of the Act requires that the SEC review the auditors of public companies no less than every three years to asses the accounting firm's compliance with the Act. The Board is to conduct inspections of registered firms annually if they audit more than 100 public companies and every three years for all others.

Audit Committee

Purpose

Section 301 of the Act requires a public company to establish an audit committee. The committee is responsible for the work of the accounting firm and any legal counsel or other independent advisors used. The accounting firm reports directly to the audit committee. The audit committee is responsible for complaints received by the company regarding the audit as well as employee concerns about questionable accounting.

Members

The audit committee is appointed by the board of directors and is comprised of only independent directors. If the board does not appoint an audit committee, then the entire board is considered the audit committee, but the requirement of independent directors still applies. A maximum or minimum number of directors is not specified for the committee, however, it does specify that at least one member must be a financial expert.

External Auditor

The audit committee is also responsible for the external auditor. The audit committee hires the auditor, compensates him and oversees the work. It also resolves any disputes between management and the outside auditor. The external auditor reports directly to the audit committee.

CEO and CFO Certification

Requirements

Section 302 requires that the financial reports of a public company be certified by the principal financial officers and the principal executive officer. They must certify that they have reviewed the report and based on their knowledge:

it does not contain any untrue statement or omit any material fact;
it fairly represents the financial condition of the company. This includes the selection and application of accounting policies, disclosure of material financial information, and the inclusion of any other information that gives the investor an

accurate and complete picture of the company.[389]

They must also certify under Section 404:

that they have established internal controls relating to the reporting of material information and that they have reported all major deficiencies and weaknesses in the internal reporting system and in the internal audit committee;

---

389 **Text of 302 Certification**

The text of the 302 Certification for a Form 10-Q Report is as follows:

[Beginning of Certification]

CERTIFICATIONS

I, [identify the certifying individual], certify that:

1. I have reviewed this quarterly report on Form 10-Q of [identify registrant];
2. Based on my knowledge, this quarterly report does not contain any untrue statement of a material fact or omit to state a material fact necessary to make the statements made, in light of the circumstances under which such statements were made, not misleading with respect to the period covered by this quarterly report;
3. Based on my knowledge, the financial statements, and other financial information included in this quarterly report, fairly present in all material respects the financial condition, results of operations and cash flows of the registrant as of, and for, the periods presented in this quarterly report;
4. The registrant's other certifying officers and I are responsible for establishing and maintaining disclosure controls and procedures (as defined in Exchange Act Rules 13a-14 and 15d-14) for the registrant and we have:
   a. designed such disclosure controls and procedures to ensure that material information relating to the registrant, including its consolidated subsidiaries, is made known to us by others within those entities, particularly during the period in which this quarterly report is being prepared;
   b. evaluated the effectiveness of the registrant's disclosure controls and procedures as of a date within 90 days prior to the filing date of this quarterly report (the "Evaluation Date"); and
   c. presented in this quarterly report our conclusions about the effectiveness of the disclosure controls and procedures based on our evaluation as of the Evaluation Date;
5. The registrant's other certifying officers and I have disclosed, based on our most recent evaluation, to the registrant's auditors and the audit committee of registrant's board of directors (or persons performing the equivalent function):
   a. all significant deficiencies in the design or operation of internal controls which could adversely affect the registrant's ability to record, process, summarize and report financial data and have identified for the registrant's auditors any material weaknesses in internal controls; and
   b. any fraud, whether or not material, that involves management or other employees who have a significant role in the registrant's internal controls; and
6. The registrant's other certifying officers and I have indicated in this quarterly report whether or not there were significant changes in internal controls or in other factors that could significantly affect internal controls subsequent to the date of our most recent evaluation, including any corrective actions with regard to significant deficiencies and material weaknesses.

Date: ....

____

[Signature]
[Title]

[End of Certification]

that they have indicated in the report any significant changes which would affect the internal controls.

The SEC considers the certification requirement to go beyond just confirming that the disclosures conform to GAAP, and is looking for assurance that the quarterly and annual reports show material accuracy and completeness.

Defense for Fraud or Misleading Financial Statement

One defense that has been used in the past with actions involving fraud or misleading financial statements has been that the defendant relied on someone else for the information. This defense may or may not still be valid with the higher standards required by the Act. As noted above, Section 404 requires the CEO and CFO to certify that internal controls have been established concerning the reporting of material information and the major weaknesses in that control system. This is to be done within 90 days prior to filing the report. The Act also authorizes the directors on the audit committee to obtain independent counsel and financial experts. If the audit committee elects not to hire the independent outside experts, then they may be deemed to be participants in alleged fraud. Financial issues should be discussed by the outside auditors, the audit committee, and management and those discussions documented.

Internal Controls

The internal control process is discussed in Section 404 of the Act. The SEC has recently acknowledged that the original attestation process would require a massive amount of work and have relaxed some of the reporting requirements. The final rules state that quarterly reports are unnecessary, but that material changes to the internal controls still do need to be reported quarterly.

The SEC leaves the internal control process to the discretion of the company, although it recommends they have a committee to consider the materiality of certain information and to determine the disclosure requirements on a timely basis.

The internal control report must include a statement concerning management's responsibility for establishing and maintaining the internal control over the financial reporting process; an assessment by management of the effectiveness of the internal control process over financial reporting; a statement explaining the framework used to evaluate the internal control process; and, any material change in the internal control process. This should include a code of ethics for senior officers.

The SEC has adopted rules requiring public companies to have disclosure controls and to evaluate them within 90 days prior to the filing of the report. The failure to maintain the disclosure controls is a violation even if it does not lead to a flawed disclosure. The officers signing the reports must also certify that they have made known all of the material deficiencies in the internal controls which would adversely affect the company's ability to report the financial data.

Oil and gas companies should implement internal controls to review the reserves reported to the public. This would include a review of the proved reserve values themselves, how they compare to prior years' numbers, the methodology used to calculate the reserves, the supporting data for the reserves, and the existence or need for an external audit by an independent reserve firm.

Oil and gas companies should have an internal review section to review the reserves reported by the business unites. It should be able to act independently and have the authority to require the business units to revise their reserve estimate if they are not in compliance with applicable reserve definitions and standards. Ideally, the reserve review section should include members from the geological, engineering, accounting and legal departments. They should have the authority to restate prior reserves if material problems are discovered. They should have the ability to hire an outside auditor if they feel it is needed.

The internal review section should examine any significant revisions to the prior years' estimate, including the delay in drilling any proved undeveloped locations or costs in excess of those estimated. They should check for the booking of reserves past the life of any concessions or any locations booked that have not been approved by the appropriate governmental agencies.

The internal audit section should insure the procedures and policies related to booking proved reserves are current and comply with the latest interpretations and guidance issued by the SEC and the relevant international countries.

Improper Influence

Influence on Auditor

Sarbanes-Oxley, in Section 303, makes it unlawful for any director, officer, or person acting under their direction to fraudulently influence any accountant doing audits for the company. This also restricts the company from pressuring the auditor to limit the scope of the audit or to not object to inappropriate accounting treatment.

Types of Improper Influence

Improper influence will cover a wide range of activities. The offer of employment or bribes is considered undue influence. Physical threats and blackmail are undue influence. Providing misleading or untrue information is also undue influence.

Code of Ethics

The Act in Section 406 requires that the companies disclose whether or not they have adopted a code of ethics for their senior financial officers. If they have not, they must state the reasons why in the disclosure.

Record Retention

Section 802 of the Act requires accounting firms to retain all relevant records for 7 years after the audit.

The rules also require the retention of documents that contain information relating to a significant matter that is inconsistent with the final conclusions of the auditor.

Penalties

Failure to Certify

Sarbanes-Oxley sets out criminal penalties for failure to certify. It also notes that certification applies not only to year end financial statements, but also to unaudited quarterly reports.

False Reports

The CEO and CFO are required to certify that the financial statement fairly represents the financial conditions and the results of operations in all material aspects. An officer who certifies the reports knowing that they are false can be subject to criminal penalties including fines of up to $1 million and imprisonment of up to 10 years. If the false certification was made willfully, the fine can be up to $5 million and 20 years imprisonment.

Restatement of Financial Condition

Under Section 304 of the Act, if the misconduct of the CEO or CFO requires the company to restate its financial condition, the CEO or CFO will be required to reimburse the company for incentive compensation and profits from the sale of securities during the 12 months period following the filing.

Altering Documents

Section 802 also provides for penalties for destruction, alteration, or falsifying records. The penalties will relate to the gravity of the alteration and range from fines to up to 20 years in prison.

Retaliation against Informants

Retaliation against an informer or witness is punishable by fine and imprisonment of up to 10 years under federal statute. Section 806 also provides protection of employees of public companies and provides a civil remedy.

Auditor Independence

Auditor independence is covered under Title II of the Act.

Partner Rotation

Section 203 calls for rotation of audit partners and specifies the time limits that an audit partner can serve in a particular position on the audit team. "Specialty partners" who include tax or valuation specialists are exempt from the partner rotation requirements. Partners consulted on technical issues and who are normally assigned office duties are also exempt from rotation requirements.

Conflicts of Interest

Section 205 provides that individuals involved in the audit cannot accept employment by the audited company in a "financial reporting oversight role" until after the next audit has been completed. This is a position where the individual has the authority to influence the contents of the financial reports.

Non-audit Services

The new SEC rules provide a list of non-audit services that the audit company can not provide to their clients.

One recent issue that has come up is the increasing reliance of boards to use outside consultants to help determine the company's top executive's compensation. The problem is that the human resource outside consultant used to provide guidance on executive compensation to the board may already be employed by the company, at a much higher total billing, to provide other human resource services. How can the human-resource companies provide impartial advice to the board when the majority of their billings come from an individual for whom they are recommending a compensation package? The SEC has not addressed this issue yet, but some have made the recommendation that consultants helping with executive compensation packages not be allowed to perform other services for the company. This is similar to the current restrictions on accountants.

Appendix A

SPE Petroleum Reserves Definitions

Reserves derived under these definitions rely on the integrity, skill, and judgment of the evaluator and are affected by the geological complexity, stage of development, degree of depletion of the reservoirs, and amount of available data. Use of these definitions should sharpen the distinction between the various classifications and provide more consistent reserves reporting.

Definitions

Reserves are those quantities of petroleum which are anticipated to be commercially recovered from known accumulations from a given date forward. All reserve estimates involve some degree of uncertainty. The uncertainty depends chiefly on the amount of reliable geologic and engineering data available at the time of the estimate and the interpretation of these data. The relative degree of uncertainty may be conveyed by placing reserves into one of two principal classifications, either proved or unproved. Unproved reserves are less certain to be recovered than proved reserves and may be further sub-classified as probable and possible reserves to denote progressively increasing uncertainty in their recoverability.

The intent of the Society of Petroleum Engineers (SPE) and World Petroleum Congress (WPC) in approving additional classifications beyond proved reserves is to facilitate consistency among professionals using such terms. In presenting these definitions, neither organization is recommending public disclosure of reserves classified as unproved. Public disclosure of the quantities classified as unproved reserves is left to the discretion of the countries or companies involved.

Estimation of reserves is done under conditions of uncertainty. The method of estimation is called deterministic if a single best estimate of reserves is made based on known geological, engineering, and economic data. The method of estimation is called probabilistic when the known geological, engineering, and economic data are used to generate a range of estimates and their associated probabilities. Identifying reserves as proved, probable, and possible has been the most frequent classification method and gives an indication of the probability of recovery. Because of potential differences in uncertainty, caution should be exercised when aggregating reserves of different classifications.

Reserves estimates will generally be revised as additional geologic or engineering data becomes available or as economic conditions change. Reserves do not include quantities of petroleum being held in inventory, and may be reduced for usage or processing losses if required for financial reporting.

Reserves may be attributed to either natural energy or improved recovery methods. Improved recovery methods include all methods for supplementing natural energy or altering natural forces in the reservoir to increase ultimate recovery. Examples of such methods are pressure maintenance, cycling, waterflooding, thermal methods, chemical flooding, and the use of miscible and immiscible displacement fluids. Other improved recovery methods may be developed in the future as petroleum technology continues to evolve.

Proved Reserves

Proved reserves are those quantities of petroleum which, by analysis of geological and engineering data, can be estimated with reasonable certainty to be commercially recoverable, from a given date forward, from known reservoirs and under current economic conditions, operating methods, and government regulations. Proved reserves can be categorized as developed or undeveloped.

If deterministic methods are used, the term reasonable certainty is intended to express a high degree of confidence that the quantities will be recovered. If probabilistic methods are used, there should be at least a 90% probability that the quantities actually recovered will equal or exceed the estimate.

Establishment of current economic conditions should include relevant historical petroleum prices and associated costs and may involve an averaging period that is consistent with the purpose of the reserve estimate, appropriate contract obligations, corporate procedures, and government regulations involved in reporting these reserves.

In general, reserves are considered proved if the commercial producibility of the reservoir is supported by actual production or formation tests. In this context, the term proved refers to the actual quantities of petroleum reserves and not just the productivity of the well or reservoir. In certain cases, proved reserves may be assigned on the basis of well logs and/or core analysis that indicate the subject reservoir is hydrocarbon bearing and is analogous to reservoirs in the same area that are producing or have demonstrated the ability to produce on formation tests.

The area of the reservoir considered as proved includes (1) the area delineated by drilling and defined by fluid contacts, if any, and (2) the undrilled portions of the reservoir that can reasonably be judged as commercially productive on the basis of available geological and engineering data. In the absence of data on fluid contacts, the lowest known occurrence of hydrocarbons controls the proved limit unless otherwise indicated by definitive geological, engineering or performance data.

Reserves may be classified as proved if facilities to process and transport those reserves to market are operational at the time of the estimate or there is a reasonable expectation that such facilities will be installed. Reserves in undeveloped locations may be classified as proved undeveloped provided (1) the locations are direct offsets to wells that have indicated commercial production in the objective formation, (2) it is reasonably certain such locations are within the known proved productive limits of the objective formation, (3) the locations conform to existing well spacing regulations where applicable, and (4) it is reasonably certain the locations will be developed. Reserves from other locations are categorized as proved undeveloped only where interpretations of geological and engineering data from wells indicate with reasonable certainty that the objective formation is laterally continuous and contains commercially recoverable petroleum at locations beyond direct offsets.

Reserves which are to be produced through the application of established improved recovery methods are included in the proved classification when (1) successful testing by a pilot project or favorable response of an installed program in the same or an analogous reservoir with similar rock and fluid properties provides support for the analysis on which the project was based, and, (2) it is reasonably certain that the project will proceed. Reserves to be recovered by improved recovery methods that have yet to be established through commercially successful applications are included in the proved classification only (1) after a favorable production response from the subject reservoir from either (a) a representative pilot or (b) an installed program where the response provides support for the analysis on which the project is based and (2) it is reasonably certain the project will proceed.

Unproved Reserves

Unproved reserves are based on geologic and/or engineering data similar to that used in estimates of proved reserves; but technical, contractual, economic, or regulatory uncertainties preclude such reserves being classified as proved. Unproved reserves may be further classified as probable reserves and possible reserves.

Unproved reserves may be estimated assuming future economic conditions different from those prevailing at the time of the estimate. The effect of possible future improvements in economic conditions and technological developments can be expressed by allocating appropriate quantities of reserves to the probable and possible classifications.

Probable Reserves

Probable reserves are those unproved reserves which analysis of geological and engineering data suggests are more likely than not to be recoverable. In this context, when probabilistic methods are used, there should be at least a 50% probability that the quantities actually recovered will equal or exceed the sum of estimated proved plus probable reserves.

In general, probable reserves may include (1) reserves anticipated to be proved by normal step-out drilling where sub-surface control is inadequate to classify these reserves as proved, (2) reserves in formations that appear to be productive based on well log characteristics but lack core data or definitive tests and which are not analogous to producing or proved reservoirs in the area, (3) incremental reserves attributable to infill drilling that could have been classified as proved if closer statutory spacing had been approved at the time of the estimate, (4) reserves attributable to improved recovery methods that have been established by repeated commercially successful applications when (a) a project or pilot is planned but not in operation and (b) rock, fluid, and reservoir characteristics appear favorable for commercial application, (5) reserves in an area of the formation that appears to be separated from the proved area by faulting and the geologic interpretation indicates the subject area is structurally higher than the proved area, (6) reserves attributable to a future workover, treatment, re-treatment, change of equipment, or other mechanical procedures, where such procedure has not been proved successful in wells which exhibit similar behavior in analogous reservoirs, and (7) incremental reserves in proved reservoirs where an alternative interpretation of performance or volumetric data indicates more reserves than can be classified as proved.

Possible Reserves

Possible reserves are those unproved reserves which analysis of geological and engineering data suggests are less likely to be recoverable than probable reserves. In this context, when probabilistic methods are used, there should be at least a 10% probability that the quantities actually recovered will equal or exceed the sum of estimated proved plus probable plus possible reserves.

In general, possible reserves may include (1) reserves which, based on geological interpretations, could possibly exist beyond areas classified as probable, (2) reserves in formations that appear to be petroleum bearing based on log and core analysis but may not be productive at commercial rates, (3) incremental reserves attributed to infill drilling that are subject to technical uncertainty, (4) reserves attributed to improved recovery methods when (a) a project or pilot is planned but not in operation and (b) rock, fluid, and reservoir characteristics are such that a reasonable doubt exists that the project will be commercial, and (5) reserves in an area of the formation that appears to be separated from the proved area by faulting and geological interpretation indicates the subject area is structurally lower than the proved area.

Reserve Status Categories

Reserve status categories define the development and producing status of wells and reservoirs.

Developed: Developed reserves are expected to be recovered from existing wells including reserves behind pipe. Improved recovery reserves are considered developed only after the necessary equipment has been installed, or when the costs to do so are relatively minor. Developed reserves may be sub-categorized as producing or non-producing.

Producing: Reserves subcategorized as producing are expected to be recovered from completion intervals which are open and producing at the time of the estimate. Improved recovery reserves are considered producing only after the improved recovery project is in operation.

Non-producing: Reserves subcategorized as non-producing include shut-in and behind-pipe reserves. Shut-in reserves are expected to be recovered from (1) completion intervals which are open at the time of the estimate but which have not started producing, (2) wells which were shut-in for market conditions or pipeline connections, or (3) wells not capable of production for mechanical reasons. Behind-pipe reserves are expected to be recovered from zones in existing wells, which will require additional completion work or future recompletion prior to the start of production.

Undeveloped Reserves: Undeveloped reserves are expected to be recovered: (1) from new wells on undrilled acreage, (2) from deepening existing wells to a different reservoir, or (3) where a relatively large expenditure is required to (a) recomplete an existing well or (b) install production or transportation facilities for primary or improved recovery projects.

Approved by the Board of Directors, Society of Petroleum Engineers (SPE) Inc., and the Executive Board, World Petroleum Congresses (WPC), March 1997

Appendix B

Sample Letter Sent to Oil and Gas Producers

The following letter was sent in February 2004 by the Division of Corporation Finance to registrants identified as being primarily engaged in the production of oil and gas. All registrants with subsidiaries or operations engaged in the production of oil and gas should consider this letter in the preparation of their filings with the Commission.

February 24, 2004

Chief Financial Officer
Company
Address

Re: Company
File No.: xxx-xxxxx

Dear Chief Financial Officer,

As a producer of oil and gas, you are subject to the disclosure requirements of FASB Statement No. 69, Disclosures about Oil and Gas Producing Activities (FAS 69). We have recently become aware of questions that have arisen with respect to the required disclosures of FAS 69 upon the adoption of FASB Statement No. 143, Accounting for Asset Retirement Obligations (FAS 143). After consideration by our staff, including discussions with the FASB staff, and to maintain comparability among oil and gas companies in preparing the FAS 69 disclosures for their 2003 annual report, we offer the following observations about the required disclosures that you should consider in preparing your Form 10-K/KSB.

Among other things, FAS 143 requires the recognition of a liability for a legal obligation associated with the retirement of a long-lived assets that results from the acquisition, construction, development, and (or) the normal operation of a long-lived asset. The initial recognition of a liability for an asset retirement obligation increases the carrying amount of the related long-lived asset by the same amount as the liability. In periods subsequent to initial measurement, period-to-period changes in the liability are recognized for the passage of time (accretion) and revisions to the original estimate of the liability. Additionally, the capitalized asset retirement cost is subsequently allocated to expense using a systematic and rational method over its useful life.

The questions raised concern how recognition of a liability for an asset retirement obligation and the related depreciation of the asset and accretion of the liability under FAS 143 impact the required disclosures under paragraphs 18 through 34 of FAS 69. We note that FAS 143 did not amend FAS 69.

FAS 69 - paragraphs 18-20, Disclosures of Capitalized Costs Relating to Oil and Gas Producing Activities (Capitalized Costs)

We believe the reported carrying value of oil and gas properties should include the related asset retirement costs and accumulated depreciation, depletion and amortization should include the accumulated allocation of the asset retirement costs since the beginning of the respective property's productive life.

The Basis of Conclusions to FAS 143 discusses the Board's conclusion regarding the capitalization of asset retirement costs by stating "a requirement for capitalization of an asset retirement cost along with a requirement for the systematic and rational allocation of it to expense achieves the objectives of (a) obtaining a measure of cost that more closely reflects the entity's total investment in the assets and (b) permitting the allocation of the cost, or portions thereof, to expense in the periods in which the related asset is expected to provide benefits." Excluding net capitalized asset retirement costs from the capitalized costs disclosure would essentially result in a presentation of capitalized costs that is not reflective of the entity's total investment in the asset, which is contrary to one of the objectives of FAS 143.

FAS 69 - paragraphs 21-23, Disclosures of Costs Incurred in Oil and Gas Property Acquisition, Exploration and Development Activities (Costs Incurred)

We believe an entity should include asset retirement costs in its Costs Incurred disclosures in the year that the liability is incurred, rather than on a cash basis.

Paragraph 21 requires an entity to disclose Costs Incurred during the year whether those costs are capitalized or charged to expense. We believe FAS 69 clearly indicates that the disclosure was intended to be on an accrual, rather than a cash, basis. Additionally, FAS 143 requires an entity to recognize the asset retirement costs and liability in the period in which it incurs the legal obligation—through the acquisition or development of an asset or through normal operation of the asset. The cost of an asset retirement obligation is not incurred when the asset is retired and the obligation is settled. Accordingly, an entity should disclose the costs associated with an asset retirement obligation in the period in which that obligation is incurred. That is, the Costs Incurred disclosures in a given period should include asset retirement costs capitalized during the year and any gains or losses recognized upon settlement of asset retirement obligations during the period.

FAS 69 - paragraphs 24-29, Disclosure of the Results of Operations for Oil and Gas Producing Activities (Results of Operations)

We believe accretion of the liability for an asset retirement obligation should be included in the Results of Operations disclosure either as a separate line item, if material, or included in the same line item as it is presented on the statement of operations.

Paragraph 14 and B57 of FAS 143 specify that the accretion expense resulting from recognition of the changes in the liability for an asset retirement obligation due to the passage of time be classified as an operating item in the statement of income. Therefore, it follows that the accretion expense related to oil and gas properties' asset retirement obligations should be included in the FAS 69 Results of Operations disclosure.

FAS 69 - Paragraphs 30-34, Disclosure of a Standardized Measure of Discounted Future Net Cash Flows Relating to Proved Oil and Gas Reserve Quantities (Standardized Measure)

We believe that an entity should include the future cash flows related to the settlement of an asset retirement obligation in its Standardized Measure disclosure.

Paragraph 30 states: "A standardized measure of discounted future net cash flows relating to an enterprise's interests in (a) proved oil and gas reserves ... and (b) oil and gas subject to purchase under long-term supply, purchase or similar agreements and contracts ... shall be disclosed as of the end of the year." We believe that the requirement to disclose "net cash flows" relating to an entity's interest in oil and gas reserves requires an entity to include the cash outflows associated with the settlement of an asset retirement obligation. Exclusion of the cash flows associated with a retirement obligation would be a departure from the required disclosure. However, an entity is not prohibited from disclosing the fact that cash flows associated with asset retirement obligations are included in its Standardized Measure disclosure as a point of emphasis.

****

Appendix C

AAPG Petroleum Resources Classification System and Definitions

Estimates derived under these definitions rely on the integrity, skill, and judgement of the evaluator and are affected by the geological complexity, stage of exploration or development, degree of depletion of the reservoirs, and amount of available data. Use of the definitions should sharpen the distinction between various classifications and provide more consistent resources reporting.

Definitions

The resource classification system is summarized in Figure 1 and the relevant definitions are given below. Elsewhere, resources have been defined as including all quantities of petroleum which are estimated to be initially-in-place; however, some users consider only the estimated recoverable portion to constitute a resource. In these definitions, the quantities estimated to be initially-in-place are defined as Total Petroleum-initially-in-place, Discovered Petroleum-initially-in-place and Undiscovered Petroleum-initially-in-place, and the recoverable portions are defined separately as Reserves, Contingent Resources and Prospective Resources. In any event, it should be understood that reserves constitute a subset of resources, being those quantities that are discovered (i.e. in known accumulations), recoverable, commercial and remaining.

TOTAL PETROLEUM-INITIALLY-IN-PLACE. Total Petroleum-initially-in-place is that quantity of petroleum which is estimated to exist originally in naturally occurring accumulations. Total Petroleum-initially-in-place is, therefore, that quantity of petroleum which is estimated, on a given date, to be contained in known accumulations, plus those quantities already produced therefrom, plus those estimated quantities in accumulations yet to be discovered.

Total Petroleum-initially-in-place may be subdivided into Discovered Petroleum-initially-in-place and Undiscovered Petroleum-initially-in-place, with Discovered Petroleum-initially-in-place being limited to known accumulations.

It is recognized that all Petroleum-initially-in-place quantities may constitute potentially recoverable resources since the estimation of the proportion which may be recoverable can be subject to significant uncertainty and will change with variations in commercial circumstances, technological developments and data availability. A portion of those quantities classified as unrecoverable may become recoverable resources in the future as commercial circumstances change, technological developments occur, or additional data are acquired.

DISCOVERED PETROLEUM-INITIALLY-IN-PLACE. Discovered Petroleum-initially-in-place is that quantity of petroleum which is estimated, on a given date, to be contained in known accumulations, plus those quantities already produced therefrom. Discovered Petroleum-initially-in-place may be subdivided into Commercial and Sub-commercial categories, with the estimated potentially recoverable portion being classified as Reserves and Contingent Resources respectively, as defined below.

RESERVES. Reserves are defined as those quantities of petroleum which are anticipated to be commercially recovered from known accumulations from a given date forward. Reference should be made to the full SPE/WPC Petroleum Reserves Definitions for the complete definitions and guidelines.

Estimated recoverable quantities from known accumulations which do not fulfil the requirement of commerciality should be classified as Contingent Resources, as defined below. The definition of commerciality for an accumulation will vary according to local conditions and circumstances and is left to the discretion of the country or company concerned. However, reserves must still be categorized according to the specific criteria of the SPE/WPC definitions and therefore proved reserves will be limited to those quantities that are commercial under current economic conditions, while probable and possible reserves may be based on future economic conditions. In general, quantities should not be classified as reserves unless there is an expectation that the accumulation will be developed and placed on production within a reasonable timeframe.

In certain circumstances, reserves may be assigned even though development may not occur for some time. An example of this would be where fields are dedicated to a long-term supply contract and will only be developed as and when they are required to satisfy that contract.

CONTINGENT RESOURCES. Contingent Resources are those quantities of petroleum which are estimated, on a given date, to be potentially recoverable from known accumulations, but which are not currently considered to be commercially recoverable.

It is recognized that some ambiguity may exist between the definitions of contingent resources and unproved reserves. This is a reflection of variations in current industry practice. It is recommended that if the degree of commitment is not such that the accumulation is expected to be developed and placed on production within a reasonable timeframe, the estimated recoverable volumes for the accumulation be classified as contingent resources.

Contingent Resources may include, for example, accumulations for which there is currently no viable market, or where commercial recovery is dependent on the development of new technology, or where evaluation of the accumulation is still at an early stage.

UNDISCOVERED PETROLEUM-INITIALLY-IN-PLACE. Undiscovered Petroleum-initially-in-place is that quantity of petroleum which is estimated, on a given date, to be contained in accumulations yet to be discovered. The estimated potentially recoverable portion of Undiscovered Petroleum-initially-in-place is classified as Prospective Resources, as defined below.

PROSPECTIVE RESOURCES. Prospective Resources are those quantities of petroleum which are estimated, on a given date, to be potentially recoverable from undiscovered accumulations.

ESTIMATED ULTIMATE RECOVERY. Estimated Ultimate Recovery (EUR) is not a resource category as such, but a term which may be applied to an individual accumulation of any status/maturity (discovered or undiscovered). Estimated Ultimate Recovery is defined as those quantities of petroleum which are estimated, on a given date, to be potentially recoverable from an accumulation, plus those quantities already produced therefrom.

AGGREGATION. Petroleum quantities classified as Reserves, Contingent Resources or Prospective Resources should not be aggregated with each other without due consideration of the significant differences in the criteria associated with their classification. In particular, there may be a significant risk that accumulations containing Contingent Resources or Prospective Resources will not achieve commercial production.

RANGE OF UNCERTAINTY. The Range of Uncertainty, as shown in Figure 1, reflects a reasonable range of estimated potentially recoverable volumes for an individual accumulation. Any estimation of resource quantities for an accumulation is subject to both technical and commercial uncertainties, and should, in general, be quoted as a range. In the case of reserves, and where appropriate, this range of uncertainty can be reflected in estimates for Proved Reserves (1P), Proved plus Probable Reserves (2P) and Proved plus Probable plus Possible Reserves (3P) scenarios. For other resource categories, the terms Low Estimate, Best Estimate and High Estimate are recommended.

The term “Best Estimate” is used here as a generic expression for the estimate considered to be the closest to the quantity that will actually be recovered from the accumulation between the date of the estimate and the time of abandonment. If probabilistic methods are used, this term would generally be a measure of central tendency of the uncertainty distribution (most likely/mode, median/P50 or mean). The terms “Low Estimate” and “High Estimate” should provide a reasonable assessment of the range of uncertainty in the Best Estimate.

For undiscovered accumulations (Prospective Resources) the range will, in general, be substantially greater than the ranges for discovered accumulations. In all cases, however, the actual range will be dependent on the amount and quality of data (both technical and commercial) which is available for that accumulation. As more data become available for a specific accumulation (e.g. additional wells, reservoir performance data) the range of uncertainty in EUR for that accumulation should be reduced.

Resources Classification System Graphical Representation

Figure 1 is a graphical representation of the definitions. The horizontal axis represents the range of uncertainty in the estimated potentially recoverable volume for an accumulation, whereas the vertical axis represents the level of status/maturity of the accumulation. Many organizations choose to further sub-divide each resource category using the vertical axis to classify accumulations on the basis of the commercial decisions required to move an accumulation towards production.

As indicated in Figure 1, the Low, Best and High Estimates of potentially recoverable volumes should reflect some comparability with the reserves categories of Proved, Proved plus Probable and Proved plus Probable plus Possible, respectively. While there may be a significant risk that sub-commercial or undiscovered accumulations will not achieve commercial production, it is useful to consider the range of potentially recoverable volumes independently of such a risk.

If probabilistic methods are used, these estimated quantities should be based on methodologies analogous to those applicable to the definitions of reserves; therefore, in general, there should be at least a 90% probability that, assuming the accumulation is developed, the quantities actually recovered will equal or exceed the Low Estimate. In addition, an equivalent probability value of 10% should, in general, be used for the High Estimate. Where deterministic methods are used, a similar analogy to the reserves definitions should be followed.

As one possible example, consider an accumulation that is currently not commercial due solely to the lack of a market. The estimated recoverable volumes are classified as Contingent Resources, with Low, Best and High estimates. Where a market is subsequently developed, and in the absence of any new technical data, the accumulation moves up into the Reserves category and the Proved Reserves estimate would be expected to approximate the previous Low Estimate.

Approved by the Board of Directors, Society of Petroleum Engineers (SPE) Inc., the Executive Board, World Petroleum Congresses (WPC), and the Executive Committee, American Association of Petroleum Geologists (AAPG), February 2000

FIGURE 1 - RESOURCES CLASSIFICATION SYSTEM

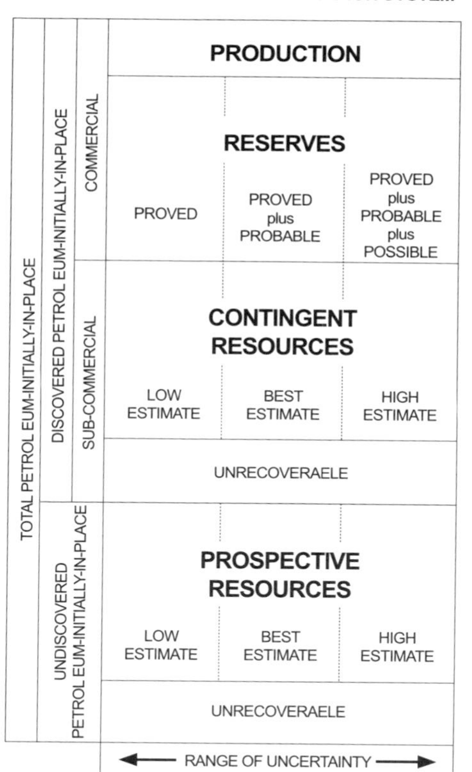

Appendix D

SEC Reserve Checklist

Quantities of Hydrocarbons

Does the reported include only volumes the SEC considers hydrocarbons?

Is any non-hydrocarbon income such as third party processing fees included in the report? The SEC does not allow any.

Reasonable Certainty

Has a conservative approach been used for reservoir parameters, decline curves and recovery factors?

Are the reserves assigned more likely to go up than go down after more data becomes available?

Economically Recoverable

Is the future net income positive?

Is there a market in place or are there negotiations to obtain one?

Existing Economic and Operating Conditions

Capital and Abandonment Cost

Has the abandonment cost been included?

Have capital costs been included for all undeveloped locations, facilities and pipelines?

Have capital costs been included for all major workovers or behind pipe reserves?

Lease Operating Cost

Have all recurring costs for the field been included?

Have the costs been allocated into fixed and variable for platforms and fields having a large infrastructure?

Has overhead cost been included?

Has COPAS been included for non operators?

Have the stated operating costs been reduced by non hydrocarbon income such as third party processing fees?

Have the costs been escalated? The SEC does not allow cost escalation.

Have costs been reduced for third party processing fees? Costs cannot be reduced for third party processing fees. They can be reduced for the amount which can be allocated to the processing of the third party product.

Taxes

Have severance taxes been included?

Have ad valorem taxes been included?

Facilities

Are facilities in place?

If not, have contracts been signed and permits obtained?

Prices

Are the prices used the year end prices or prices as of the report date?

Have the prices been based on an average of the last 12 first day of the month prices?

Have the prices been escalated? No escalation for SEC prices.

If there is a contract, have the contract prices and escalations been used?

If gas is not subject to a sales agreement or contract, has the price been based on similar gas sales?

Have alternative pricing cash flows been included only in the Disclosure Section?

Miscellaneous Issues

Flow tests

Is economic production based on actual production?

Is economic production based on flow tests?

Is economic production based on logs, cores and analogy?

Data

Has data after the report date been used? Data after the report date is not allowed.

Ownership

Are the reserves reported net to the company's interest?

Are reserves subject to a PSA reported separately from owned mineral interests?

Have Net Profits Interests been included in the lease operating expense?

Has the portion of the plant volume allocated to the lease been assigned?

Foreign Concessions

Is there an approved plan of development?

Is a market available?

Does the reserve life extend past the PSC? Reserves beyond the life of the PSC cannot be reported.

Recovery Factors

Is data available showing how the recovery factor has been determined?

If the drive mechanism is not known, has the most conservative drive mechanism been assumed?

Production Rates

Are future production rates based on decline curves, analogy, or tests?

Has the structural position of the well been considered before a future rate was assigned?

Has the capacity of the production facilities and pipelines been considered?

Has the effect of market fluctuations been considered?

Has the drive mechanism been considered?

Has the geologic area and type of rock been considered?

Condensate and GOR Forecasts

Are future yields and GOR based on decline curves or tests?

Are the future yields and GORs consistent with the drive mechanism being used?

Simulation

Is there a good history match?

Reserve Determination

Deterministic or Probabilistic

If probabilistic methods are being used, are the SEC reservoir limits being used?

Volumetric Reserves

Reservoir Limits

Are Lowest Known Hydrocarbon Limits used if there is no water contact in the well?

Is the Lowest Know Hydrocarbon based on wireline log data?

Have reservoir limits been based on seismic? The SEC does not allow downdip limits based on seismic data alone.

Effective Thickness

Has the pay been netted so that only reservoir rock is counted?

Has the mapped thickness been thickened to more than seen in the wellbore? The mapped net pay should not be greater than the net sand seen in any wellbore, even if it appears to thicken on seismic.

Has pay been thickened based on seismic?

Porosity and Water Saturation

Have accepted methods been used to calculate the porosity and water saturation?

Has the porosity calculated from logs been calibrated against the core porosity?

Have porosity and water saturation cutoffs been estimated either based on tests or analogy?

Recovery Factors

Has the basis for the recovery factor used been documented?

Is the recovery factor based on analogy?

Is the drive mechanism known?

If the drive mechanism is not known, has the most conservative drive been estimated? In general, this is a water drive recovery factor for gas and a depletion drive recovery factor for oil.

Undeveloped Reserves

Have more than 8 direct offsets been assigned as direct offsets to a commercial well? The SEC allows only one location offset to be assigned as proved without reasonable certainty of reservoir continuity.

If more than 8 offsets have been assigned, can reservoir continuity be shown between the commercial wells? Reservoir continuity must be based on pressure information.

Are any of the locations outside the reservoir boundaries or below the Lowest Known Hydrocarbon? All locations should be within the SEC defined limits of the reservoir.

Are the locations legal locations?

Do the locations go beyond the life of the PSA? Reserves beyond the life of the PSC should not be included.

Is there a commitment by the operator to drill?

Is a development plan in place?

Does the plan of development extend for more than 5 years?

Has the plan of development been approved by the host government?

Has simulation been used to estimate the undeveloped reserves? Simulation results should not be used alone to document SEC proved reserves.

Is there documentation to support the reserves assigned?

EOR Projects

Is there a successful pilot project or is the project fully installed before proved developed reserves are assigned?

Have proved developed reserves been assigned to more of the reservoir than is seeing a response?

For undeveloped EOR reserves, is there an analogy in the area in the same reservoir?

What is the basis for the projected increase in reserves?

Are facilities in place or been committed to?

Appendix E

Data Requirements for Reserve Estimates

General

- Type of report required – SEC, SPE etc.
- Reserve categories to be included in the report
- As of date for the report
- Deadline for report
- Number of copies of the report required
- Base map showing all current wells and proposed locations
- Well list giving the well names and coordinates for each well and location

Ownership

- Type of ownership – mineral or PSC
- Expiration date of concessions
- Interests for costs and revenues including reversions and net profit interests
- Payout data for reversions

Costs, Prices and Taxes

- Monthly operating cost data for the last year
- Ad valorem tax data
- Capital cost data for drilling, workovers, etc.
- AFE's document capital costs
- Operator's budget and plans to drill or workover wells
- Price data for all products, including transportation, BTU, differentials
- Transportation, processing and any other applicable fees
- Pricing differentials
- Fuel and shrinkage volumes

• Any production tax exemptions

Volumetric

• Logs – porosity and resistivity (including digital data)
• Field logs and offset logs should be included
• Directional surveys
• Client petrophysical interpretations
• Structure and isopach maps
• Cores – whole on key wells, sidewall on others
• Seismic data to define faults and other major features
• Velocity surveys
• Pressure Information, including bottom hole pressures and shut-in pressures
• Bottom Hole temperature information
• Type of drive mechanism from analogy
• Flow tests
• Drill stem tests or MDT's
• Fluid samples

Performance

• Monthly production for oil, gas and water including cumulative for each well. If lease or total field production is reported, include the number of producing wells.
• Liquid gravity
• BTU and shrinkage
• Plant product yields
• PVT data
• Monthly test data including rates for oil, gas and water and pressures – flowing and shut-in
• Bottom hole pressure data – current and historical
• SITP and FTP information
• Historical data on well work including all recompletions, artificial lift, and

compression

- All perforations – current and historical
- Injections volumes and pressures
- Information on any new drilling including testing, and completion reports
- Recompletion and workover reports
- Future development plans
- Future recompletions and drilling with timing and costs
- Simulation studies

Appendix F

China Petroleum Resources/Reserves Classification

GAO Ruiqi[1], LU Minggang[1], ZHA Quanheng[1], XIAO Deming[1], HU Yundong[1,2]

1 China Petroleum Reserves Office, Ministry of Land & Resources

2 China University of Geosciences at Beijing

Comments are welcome forwarded to:

Att: Mr. **Hu Yundong**

Petroleum Reserves Office (PRO)

PetroChina Research Institute

20 Xueyuan Rd., Beijing 100083

P.R.China

E-mail: hyd@petrochina.com.cn or dony_129@sina.com

Tel & Fax: 86 10 62097426

Scope

The standard stipulates the classifications and definitions of the petroleum resources / reserves (hereinafter referred to as resources / reserves).

The standard is applicable to estimation, auditing and statistics of resources / reserves, also applicable to the approvals of domestic concession and development plan, transfer of mineral ownership and the appraisal of property by third party in the financing activities during petroleum exploration and development.

Terms and definitions

The following terms and definitions are applicable to the standard.

2.1 ***Initially-in-place Volumes:*** generally, the initially-in-place volumes are the natural accumulation quantities of oil and gas resulted from geological process in the crust. In other words, they are the volumes of hydrocarbon and related substances existed in a reservoir before any volume has been produced and are expressed in the measurement of surface standard condition (20o℃, 0.101MPa).

2.2 ***recoverable Volumes:*** recoverable volumes are the expected or estimated recoverable parts of petroleum initially in place.

2.3 ***Resources:*** resources are a general designation of initially-in-place volumes and recoverable volumes.

2.4 ***Reserves:*** reserves are a general designation of discovered resources which are particularly called *Geological reserves and Recoverable reserves*, respectively refer to the discovered initially-in-place volumes and recoverable volumes. The recoverable reserves refer to *Technically estimated ultimate recovery* and *Economic initially recoverable reserves*.

2.5 ***Technically Estimated Ultimate Recoveries*** **(TEUR)**: technically estimated ultimate recoveries are those quantities of petroleum which are theoretically or analogically estimated to be recoverable from discovered accumulations under given technological condition.

2.6 ***Economic Initially Recoverable Reserves*** **(EIRR)**: economic initially recoverable reserves are those quantities of petroleum which are anticipated to be economically recoverable from discovered accumulations under existing economic conditions (such as prices, costs, *etc.*) and under current executed or planned to be established technical operating conditions.

2.7 ***Residual Unrecoverable Volumes*** **(RUV)**: residual unrecoverable volumes are the differences between initially-in-place volumes and recoverable volumes.

Exploration and development phases

Petroleum exploration and development is a progressive process. The complete exploration and development process can be divided into five phases: regional reconnaissance, general exploration, reservoir appraisal, development construction and production operation.

3.1 ***Regional Reconnaissance:*** in the phase, the regional geology survey is made in the basin, depression and their surrounding areas, then selective seismic and non-seismic reconnaissance surveys are made, regional reconnaissance well is drilled in order to understand the basic petroleum geological conditions about hydrocarbon source rocks, potential reservoirs and seals etc. and the potential plays or favorable hydrocarbon bearing zones are delineated.

3.2 ***General Exploration:*** the detailed seismic and other necessary geochemical surveys are made in the favorable hydrocarbon bearing zones in order to ascertain the traps and their distribution, then the favorable traps are selected to drill the preliminary prospecting well in order to identify the basic features of structure, reservoir and seal, finally the oil and/or gas reservoirs/fields are discovered with the understanding of basic characteristics of the reservoirs.

3.3 ***Reservoir Appraisal:*** after the oil and/or gas is discovered in the general exploration phase, the following work should be done progressively in order to develop the reservoirs scientifically and economically, which includes detailed seismic survey or 3D seismic exploration and appraisal well(s) drilling for the purpose of identifying structure configuration, fault location, reservoir distribution and reservoir properties variation, identifying reservoir type, pore morphology, drive mechanism, fluid properties, distributions and productivities, and understanding the recovery operation conditions and economics. Then the development plan is carried out.

3.4 ***Development Construction:*** production wells are drilled according to the development plan, the construction of associated facilities is fulfilled, necessary additional data are acquired, oil and gas reserves are reviewed and productivities are re-examined. The reservoirs are ready for production.

3.5 ***Production Operation:*** planned oil and gas production in the blocks/fields is operated with fulfilled facility construction, and necessary production adjustment, reconstructing or improvement are made in appropriate time in order to improve recovery, reasonably utilize the oil and gas resources and enhance the economic benefit.

Reserves Economic Criteria

The economic viabilities for an oil and gas reservoir/field development are judged through the feasibility study in different exploration and development phases. Generally, the reserve economic criteria are divided into three categories: economic, sub-economic and undetermined economic.

4.1 ***Economic***: based on the market condition of the time, i.e. oil and gas prices and development costs at the time of the reserve estimation, oil and gas production is believed technically feasible with the other conditions allowable, such as environment, etc. The economic viability refers to the reserves revenue can return the investment.

4.2 ***Sub-economic***: based on the market condition of the time, oil and gas production is not economic, but in the projected feasible market condition or in the conjectural probably occurring market condition, or in the condition with the investment environment improved, the production would be economic.

4.3 ***Undetermined Economic***: only the general geological study is made for the reservoirs/fields. Since there are only some preliminary assumptions with many uncertainties existed for the reservoir complexity, reservoir scale, development technology applications and future market, it is impossible to determine the economic viability is economic or sub-economic.

Resources /Reserves Classification

Resources/reserves classification is made based on the phases of exploration and development, the degree of geologic certainty and productivity confirmation.

5.1 Classification Framework Classification framework is shown in FIGURE. 1. (P. 112)

5.2 Initially-in-place Volumes Classification

5.2.1 ***Total Petroleum Initially-in-place Volumes*** **(TPIIP)**: total petroleum initially-in-place volumes are the total initially existed oil and gas in known and unknown accumulations which are estimated by adopting the pertinent method, based on the geological, geophysical and laboratory data provided in different exploration and development phases.

5.2.2 ***Geological Reserves*** **(GR)**: geological reserves (i.e. discovered petroleum initially-in-place volumes) are the total oil and gas quantities estimated based on seismic, drilling, well logging and test data in the known reservoirs/fields after the oil and gas was found by drilling. Geological reserves are subdivided into three categories: *Measured, Indicated and Inferred.*

5.2.2.1 ***Measured Geological Reserves*** **(MEGR)**: measured geological reserves are estimated with a high level of confidence and relative error not more than ±20 after the reservoirs have been proved economically recoverable by appraisal drilling at appraisal phase. The estimation of measured geological reserves should identifying the reservoir type, pore morphology, drive mechanism, fluid properties, distributions and productivities, *etc*. Fluid contacts or the lowest known hydrocarbon should be determined by drilling, logging and test data or reliable pressure data. The reasonable well spacing or primary development well pattern should be used in the delineation of measured limits. All parameters in the volumetric approach should have a high degree of certainty.

5.2.2.2 ***Indicated Geological Reserves*** **(IDGR)**: indicated geological reserves are estimated with a moderate level of confidence and relative error not more than ±50 when oil and/or gas economic flow is obtained from prospecting well at general exploration phase. The estimation of indicated geological reserves should preliminarily ascertaining structure configuration, formation continuity, oil and gas distribution, reservoir type, fluid properties and productivities, etc. The geological confidence degree is moderate, which can be as evidence for drilling reservoir appraisal wells, making conceptual design or development plan.

5.2.2.3 ***Inferred Geological Reserves*** **(IFGR)**: inferred geological reserves are estimated with a rather low level of confidence in the probably existing oil and/or gas reservoirs with certain degree of exploration potentiality at general exploration phase when oil and/or gas flow is obtained from preliminary prospecting well or in the case that the integrative interpretation indicates that there is probably oil and/or gas layers existed. The estimation of the inferred geological reserves requires basically understanding the structure configurations and reservoir conditions.

5.2.3 ***Undiscovered Petroleum Initially-in-place Volumes*** **(UPIIP)**: undiscovered petroleum initially-in- place volumes are the quantities of the total oil and gas estimated in unknown accumulations based on prediction, which are categorized as *Petroleum initially-in-place in prospects* and *Unmapped petroleum initially-in-place.*

5.2.3.1 ***Petroleum Initially-in-place In Prospects*** **(PIIPIP)**: which is referred to the total petroleum initially-in-place in the known favorable traps or blocks/ formations adjacent to oil and/or gas field estimated by trap method at the early stage of general exploration phase. The estimation is based on the analysis and analogy of petroleum geological conditions.

5.2.3.2 ***Unmapped Petroleum Initially-in-place*** **(UMPIIP)**: which is referred to the total petroleum in place estimated in the phase of regional reconnaissance or other exploration phase in the prospecting basin, depression or sags and belts. The estimation is based on geological, geophysical and geochemical reconnaissance, regional exploratory well data. *Unmapped petroleum initially-in-place* generally is the result of subtracting the difference of *the Geological reserves and the Petroleum initially-in-place in prospects from the Total petroleum initially-in-place.*

5.3 Recoverable Volume Classifications

Recoverable volumes are divided into *Recoverable reserves* and *Recoverable resources.*

5.3.1 ***Recoverable Reserves*** **(RR)**: recoverable reserves are the recoverable oil and/or gas quantities from *Geological reserves*. Based on the degree of geological confidence and economic viabilities there are seven classifications (Note the inferred geological reserves are undetermined economic, which is not applicable to the economic recoverable reserves).

5.3.1.1 ***Proved Technically Estimated Ultimate Recoveries*** **(PVTEUR)**: which are the Technically estimated ultimate recoveries meeting the following requirements:

The technology (including oil and/or gas recovery technology and enhanced recovery technology, the same as hereinafter) has been operated or planned to be operated in the near future;

Already have conceptual design or development plan and which have been carried out or will be carried out in the near future;

Based on the recent average prices and costs, the feasibility study indicates that the development is economic or sub-economic.

5.3.1.2 ***Proved Economic Initially Recoverable Reserves*** **(PVEIRR)**: which are the Economic initially recoverable reserves meeting the following requirements:

Based on the different requirements, the prices and costs are on the as of date or stipulated in the contracts or agreements, and other related economic conditions are observed;

The technology has been operated or the technology demonstrated by pilot project to be favorable for operation, or the technology which is actually successful in the analogous oil and/or gas field and assured to be installed;

Already have development plan and which will be carried out in the near future; for gas there should have existing gas pipelines or gas pipeline construction agreement, and also have sales contract or agreement;

The reserve boundaries are on the fluid contacts confirmed by drilling or reliable pressure test data, or the lowest known hydrocarbon in the encountered well, and within the boundaries a reasonable well control is fulfilled;

The economic productivity is supported by either actual production or conclusive test, or the economic productivity in the objective formation is confirmed to be similar to the same formation in the wells located beyond direct offsets or the similar formation in the same well which have indicated economic production;

Feasibility study shows the development is economic;

There should be at least 80% probability that the quantities actual recovered in the future will equal or exceed the estimated *Economic initially recoverable reserves*.

5.3.1.3 ***Proved Sub-economic Initially Recoverable Reserves*** **(PVSEIRR)**: which are the differences between the Proved technically estimated ultimate recoveries and the Proved economic initially recoverable reserves, which includes two parts:

Those *Proved Technically Estimated Ultimate Recoveries* that the feasibility study indicates the development is sub-economic;

Those *Proved Technically Estimated Ultimate Recoveries* anticipated be economic but the uncertainties of contractual and/or technical in enhancing recoveries preclude such volumes being classified as *Proved Economic Initially Recoverable Reserves*.

5.3.1.4 ***Probable Technically Estimated Ultimate Recoveries*** **(PBTEUR)**: which are the Technically estimated ultimate recoveries meeting the following requirements:

Presume the probably executed operation technology;

The feasibility study shows the development is above sub-economic.

5.3.1.5 ***Probable Economic Initially Recoverable Reserves*** **(PBEIRR)**:

which are the Economic initially recoverable reserves meeting the following requirements:

Feasibility study shows the development is economic;

There should be at least 50% probability that the quantities actual recovered in the future will equal or exceed the estimated *Economic initially recoverable reserves*.

5.3.1.6 ***Probable Sub-economic Initially Recoverable Reserves*** **(PBSEIRR)**: which are the differences between the probable technically estimated ultimate recoveries and the probable economic initially recoverable reserves.

5.3.1.7 ***Possible Technically Estimated Ultimate Recoveries*** **(PSTEUR)**: which are the Technically estimated ultimate recoveries meeting the following requirements:

optimistically presume the probably adopted operation technology;

There should be at least 10% probability that the quantities actual recovered in the future will equal or exceed the estimated *Economic initially recoverable reserves*.

5.3.2 **Recoverable Resources**: Recoverable resources are the recoverable petroleum quantities from petroleum initially-in-place, which are categorized into *Recoverable resources in prospects* and *Unmapped recoverable resources*, and the recovery factors are estimated by empirical analogy.

5.3.2.1 ***Recoverable Resources in Prospects*** **(RRIP)**: which are referred to the petroleum quantities recovered from *Petroleum initially-in-place in prospects*.

5.3.2.2 ***Unmapped Recovery Resources*** **(URR)**: which are referred to the petroleum quantities recovered from *Unmapped petroleum initially-in-place*.

5.4 Reserve Status Categories

Reserve status categories define the development and producing status of proved economic recoverable reserves in two categories: *Developed and Undeveloped.*

5.4.1 ***Proved developed*** **(PD)**: it is referred to the reserves fully put into production after the completing the oil and/or gas reservoir development well pattern drilling and associated facility installment. When the facilities required by the improved recovery technology (such as water flooding *etc.*) have been established and put into production, the corresponding increased recoverable reserves are also categorized as proved developed recoverable reserves. The proved developed initially recoverable reserves are the evidences for the production analysis, adjusting and management, and are also the measurement for different level recoverable reserves accuracy comparison. Proved developed reserves should be updated regularly during the development and production. The proved developed initially recoverable reserves after subtracting accumulated recoveries are the proved developed remaining recoverable reserves.

5.4.2 ***Proved undeveloped*** **(PUD)**: it is referred to the recoverable reserves in the oil and/or gas reservoirs which have completed appraisal drilling or have a pilot production project but the production well pattern is not fulfilled.

Appendix G

Reserve Comparison Table

	U.S. SEC Reg. S-X	SEC January 2009	SPE/WPC	PRC
Purpose	Securities reporting	Securities reporting and investor understanding or oil and gas reserves	Consistent reserve definitions	Reserve Definitions of China for reporting and planning.
Reserves		Reserves are estimated remaining quantities of oil and gas and related substances anticipated to be economically producible, as of a given date, by application of development projects to known accumulations.	…those quantities of petroleum anticipated to be commercially recoverable by application of development projects to known accumulations from a given date forward under defined conditions. Reserves must further satisfy four criteria: they must be discovered, recoverable, commercial, and remaining (as of the evaluation date) based on the development project(s) applied..	Reserves are a general designation of discovered resources which are particularly called *Geological reserves* and *Recoverable reserves*, respectively refer to the discovered initially-in-place volumes and recoverable volumes. The recoverable reserves refer to *Technically estimated ultimate recovery* and *Economic initially recoverable reserves.*
Proved Categories	Developed, Undeveloped	Developed, Undeveloped	Developed Producing Shut-in Behind Pipe Undeveloped	Developed Undeveloped Basic Proved Reserves
Proved Definition	Proved oil and gas reserves are the estimated quantities of crude oil, natural gas, and natural gas liquids which geological and engineering data demonstrate with **reasonable certainty** to be recoverable in future years from known reservoirs	Proved oil and gas reserves are those quantities of oil and gas, which, by analysis of geoscience and engineering data, can be estimated with reasonable certainty to be economically producible—from a given date forward, from known reservoirs, and under existing economic conditions, operating methods, and government regulations—prior to the	Proved reserves are those quantities of petroleum which, by analysis of geological and engineering data, can be estimated with **reasonable certainty** to be commercially recoverable, from a given date forward, from known reservoirs and under current economic conditions, operating methods, and government regulations. Proved reserves can be	Proved reserves are estimated after completion or near the completion of the evaluation drilling. They are defined "Under the present technical and economic conditions, it is **reliable** reserve for recovery and social economic profit."

继表

	U.S. SEC Reg. S-X	SEC January 2009	SPE/WPC	PRC
Proved Definition	under existing economic and operating conditions, i.e., prices and costs as of the date the estimate is made.	time at which contracts providing the right to operate expire, unless evidence indicates that renewal is reasonably certain, regardless of whether deterministic or probabilistic methods are used for the estimation. The project to extract the hydrocarbons must have commenced or the operator must be reasonably certain that it will commence the project within a reasonable time.	categorized as developed or undeveloped.	
Proved Developed	Proved developed oil and gas reserves are reserves that can be expected to be recovered through existing wells with existing equipment and operating methods. Additional oil and gas expected to be obtained through the application of fluid injection or other improved recovery techniques for supplementing the natural forces and mechanisms of primary recovery should be included as "proved developed reserves" only after testing by a pilot project or after the operation of an installed program has confirmed through production response that increased recovery will be achieved.	Developed oil and gas reserves. Developed oil and gas reserves are reserves of any category that can be expected to be recovered: (i) Through existing wells with existing equipment and operating methods or in which the cost of the required equipment is relatively minor compared to the cost of a new well; and (ii) Through installed extraction equipment and infrastructure operational at the time of the reserves estimate if the extraction is by means not involving a well.	Developed reserves are expected to be recovered from existing wells including reserves behind pipe. Improved recovery reserves are considered developed only after the necessary equipment has been installed, or when the costs to do so are relatively minor. Developed reserves may be sub-categorized as producing or non-producing.	Proved developed reserves (I or A) are described in the Petroleum Reserve Standard as reserves that are identified as available under current technical and economic conditions, a development plan has been put into effect, all production wells have been completed, all facilities are in place and the reserves are on production. These reserves can be increased after facilities for improved recovery are installed. These reserves are used as the basis for development analysis and management.

继表

	U.S. SEC Reg. S-X	SEC January 2009	SPE/WPC	PRC
Proved Undeveloped	Proved undeveloped oil and gas reserves are reserves that are expected to be recovered from new wells on undrilled acreage, or from existing wells where a relatively major expenditure is required for recompletion. Reserves on undrilled acreage shall be limited to those drilling units offsetting productive units that are reasonably certain of production when drilled. Proved reserves for other undrilled units can be claimed only where it can be demonstrated with certainty that there is continuity of production from the existing productive formation. Under no circumstances should estimates, for proved undeveloped reserves be attributable to any acreage for which an application of fluid injection or other improved recovery technique is contemplated, unless such techniques have been proved effective by actual tests in the area and in the same reservoir.	Undeveloped oil and gas reserves. Undeveloped oil and gas reserves are reserves of any category that are expected to be recovered from new wells on undrilled acreage, or from existing wells where a relatively major expenditure is required for recompletion. Reserves on undrilled acreage shall be limited to those directly offsetting development spacing areas that are reasonably certain of production when drilled, unless evidence using reliable technology exists that establishes reasonable certainty of economic producibility at greater distances. Undrilled locations can be classified as having undeveloped reserves only if a development plan has been adopted indicating that they are scheduled to be drilled within five years, unless the specific circumstances, justify a longer time.) Under no circumstances shall estimates for undeveloped reserves be attributable to any acreage for which an application of fluid injection or other improved recovery technique is contemplated, unless such techniques have been proved effective by actual projects in the same reservoir or an analogous reservoir, as defined in paragraph (a)(2) of this section, or by other evidence using reliable technology establishing reasonable certainty.	Undeveloped Reserves: Undeveloped reserves are expected to be recovered: (1) from new wells on undrilled acreage, (2) from deepening existing wells to a different reservoir, or (3) where a relatively large expenditure is required to (a) recomplete an existing well or (b) install production or transportation facilities for primary or improved recovery projects. There should be an expectation that the project will be economic and that the entity has committed to implement the project in a reasonable timeframe (generally within five years; further delays should be clearly justified). Incremental recoveries through improved recovery methods that have yet to be established through routine, commercially successful applications are included as Reserves only after a favorable production response from the subject reservoir from either (a) a representative pilot or (b) an installed program, where the response provides support for the analysis on which the project is based.	The Petroleum Reserve Standard defines proved undeveloped reserves (II or B) as reserves based on "reliable reservoir parameters". They are the basis for the development plan and facilities construction. The error in these reserves should be no more than plus or minus 20%.

继表

	U.S. SEC Reg. S-X	SEC January 2009	SPE/WPC	PRC
Certainty of Proved Reserves	Reasonable certainty; certainty for undeveloped locations	Reasonable certainty	Reasonable certainty	Reliable
Pricing	Price actually received on report date. Prices can include changes provided for by contract, but not escalations based on future conditions.	Existing economic conditions include prices and costs at which economic producibility from a reservoir is to be determined. The price shall be the average price during the 12-month period prior to the ending date of the period covered by the report, determined as an unweighted arithmetic average of the first-day-of-the-month price for each month within such period, unless prices are defined by contractual arrangements, excluding escalations based upon future conditions.	Can use average prices. Current conditions for proved, but can estimate unproved reserves assuming different future economics.	Economics are not an issue other than industrial flow.
Reservoir Limits	Reservoir limits based on hydrocarbon limits or the lowest known hydrocarbon based on log data.	In the absence of data on fluid contacts, proved quantities in a reservoir are limited by the lowest known hydrocarbons (LKH) as seen in a well penetration unless geoscience, engineering, or performance data and reliable technology establishes a lower contact with reasonable certainty.	Reservoir limits based on hydrocarbon limits or the lowest known hydrocarbon based on log or test data.	The hydrocarbon bearing area is delineated based on a structure map on the top of the reservoir using logs, seismic and tests. Proved reservoir limits can be delineated based on boundary tests or by pressure tests if there is not a water contact China allows booking proved reserves below the low known hydrocarbons. The China standard is one half the distance from the low known hydrocarbon to the next wet sand.
Ownership	Company should report volumes net to their interests.	Company should report volumes net to their interests.	Not addressed.	Owned by the state.
Use of Data after Year-end	Cannot use data after the first of the year. Can footnote in the disclosure section if material.	Cannot use data after the first of the year.	Not addressed.	Not discussed.

继表

	U.S. SEC Reg. S-X	SEC January 2009	SPE/WPC	PRC
Undeveloped Locations	One offset as proved. For more than one offset, must have "**certainty**" of reservoir continuity.	Reserves on undrilled acreage shall be limited to those directly offsetting development spacing areas that are reasonably certain of production when drilled, unless evidence using reliable technology exists that establishes **reasonable certainty** of economic producibility at greater distances.	Offsets more than one legal location away are allowed if the geological and engineering data indicate with "**reasonable certainty**" that the reservoir is continuous and has economic reserves more than one legal location from production.	Proved undeveloped reserves (II or B) as reserves based on "reliable reservoir parameters". They are the basis for the development plan and facilities construction. The error in these reserves should be no more than plus or minus 20%.
Improved Recovery	Favorable response to a pilot or analogy.	Reserves which can be produced economically through application of improved recovery techniques (including, but not limited to, fluid injection) are included in the proved classification when: (A) Successful testing by a pilot project in an area of the reservoir with properties no more favorable than in the reservoir as a whole, the operation of an installed program in the reservoir or an analogous reservoir, or other evidence using reliable technology establishes the reasonable certainty of the engineering analysis on which the project or program was based; and (B) The project has been approved for development by all necessary parties and entities, including governmental entities.	Favorable response to a pilot or by analogy and reasonable certainty that the project will proceed.	Not addressed.
Market Availability	Market must be available.	There must exist, or there must be a reasonable expectation that there will exist, the legal right to produce or a revenue interest in the production, installed means of delivering oil and gas or related substances to market, and all permits and financing required to implement the project.	Reserves may be classified as proved if facilities to process and transport those reserves to market are operational at the time of the estimate or there is a reasonable expectation that such facilities will be installed.	Not discussed.

继表

	U.S. SEC Reg. S-X	SEC January 2009	SPE/WPC	PRC
Requirement for Testing	Test required in new areas or new reservoirs. In certain instances, proved reserves may be assigned to reservoirs on the basis of a combination of electrical and other type logs and core analyses which indicate the reservoirs are analogous to similar reservoirs in the same field which are producing or have demonstrated the ability to produce on a formation test.	Actual production or flow tests are no longer required if other reliable technology is used to demonstrate the reservoir is economically producible.	In general, reserves are considered proved if the commercial producibility of the reservoir is supported by actual production or formation tests. In certain cases, proved reserves may be assigned on the basis of well logs and/or core analysis that indicate the subject reservoir is hydrocarbon bearing and is analogous to reservoirs in the same area that are producing or have demonstrated the ability to produce on formation tests.	China has established minimum rates for a well to be considered or to have industrial flow.
Positive Cash Flow	Required.	The term economically producible, as it relates to a resource, means a resource which generates revenue that exceeds, or is reasonably expected to exceed, the costs of the operation. The value of the products that generate revenue shall be determined at the terminal point of oil and gas producing activities.	Requires reserves to be “commercially recoverable”.	Positive cash flow not emphasized, but must have industrial flow.
Operating Costs	Costs on date of report.	Existing economic conditions include prices and costs at which economic producibility from a reservoir is to be determined. The price shall be the average price during the 12-month period prior to the ending date of the period covered by the report, determined as an unweighted arithmetic average of the first-day-of-the-month price for each month within such period, unless prices are defined by contractual arrangements, excluding escalations based upon future conditions.	PRMS 3.1.2: The economic evaluation underlying the investment decision is based on the entity's reasonable forecast of future conditions, including costs and prices, which will exist during the life of the project (forecast case). Such forecasts are based on projected changes to current conditions; the SPE defines current conditions as the average of those existing during the previous 12 months. Alternative economic scenarios are considered in the decision process and, in some cases, to supplement reporting requirements.	Not discussed.